XI International Conference on Aerospace Education and Staffing for High-Tech Enterprises (AESHE 2021)

SHS Web of Conferences Volume 137 (2022)

Alushta, Crimea
21 – 26 September 2021

Editors:

Dmitriy A. Kozorez
Vladimir T. Kalugin
Shuai Yong
Natalia S. Severina

ISBN: 978-1-7138-4840-0

Printed from e-media with permission by:

Curran Associates, Inc.
57 Morehouse Lane
Red Hook, NY 12571

Some format issues inherent in the e-media version may also appear in this print version.

No changes have been made to the content of these proceedings. There may be changes to pagination and minor adjustments for aesthetics.

Printed with permission by Curran Associates, Inc. (2023)

For additional information, please contact EDP Sciences – Web of Conferences at the address below.

EDP Sciences – Web of Conferences
17, Avenue du Hoggar
Parc d'Activité de Courtabœuf
BP 112
F-91944 Les Ulis Cedex A
France

Phone: +33 (0) 1 69 18 75 75
Fax: +33 (0) 1 69 28 84 91

contact-edps@webofconferences.org

Additional copies of this publication are available from:

Curran Associates, Inc.
57 Morehouse Lane
Red Hook, NY 12571 USA
Phone: 845-758-0400
Fax: 845-758-2633
Email: curran@proceedings.com
Web: www.proceedings.com

TABLE OF CONTENTS

SHS Web of Conferences **137**, 00001 (2022) https://doi.org/10.1051/shsconf/202213700001
AESHE 2021

XI International Conference on
Aerospace Education and Staffing
for High-Tech Enterprises

AESHE 2021

September 21-26, 2021

Preface

This issue comprises selected works presented on the Eleventh International Conference on Aerospace Education and Staffing for High-Tech Enterprises (AESHE 2021) held in September 21-26, 2021. The Conference was hosted by the Moscow Aviation Institute (national research university) with cooperation of the Bauman Moscow State Technical University (national research university), the Association of Sino-Russian Technical Universities (ASRTU), the Ministry of Science and Higher Education of the Russian Federation.

The first time Conference was organized by Moscow Aviation Institute in 2008 at the health and training center "Alushta" Moscow Aviation Institute. The Conference is attended by leading methodologists working in aerospace education: vice-rectors, deans and heads of aerospace departments, as well as representatives of employers. The sessions discuss issues of the structure of education, the regulatory educational activities framework and new technologies impact. There is an experience exchange in organizing methodological and educational work at universities. One of the constantly discussed topics is personnel training for high-tech enterprises.

This Conference is devoted to the study of the problems associated with the development and improvement of such training. Scientific areas of the Conference: the impact of new technologies on the development of aerospace education, design of educational programs in the framework of international cooperation of universities, professional orientation of applicants for admission to aerospace universities, digital technologies in aerospace education, educational platforms in the educational process, training of personnel with higher education on an interdisciplinary basis, training of personnel in the framework of additional professional education.

Editors-in-Chief:

Dr. Dmitriy A. Kozorez, MAI, Russia
E-mail: kozorezda@mai.ru

Dr. Vladimir T. Kalugin, BMSTU, Russia
E-mail: kaluginvt@bmstu.ru

Guest Editors:

Dr. Shuai Yong, Harbin Institute of Technology, P. R. China
E-mail: zhangwei1990_hit@163.com

Dr. Natalia S. Severina, MAI, Russia
Email: severina@mai.ru

edp sciences

SHS Web of Conferences

Statement of Peer review

In submitting conference proceedings to SHS Web of Conferences, I certify to the Publisher that I adhere to the **Policy on Publishing Integrity** of the journal in order to safeguard good scientific practice in publishing.

1. All articles have been subjected to peer review administered by the proceedings editors.
2. Reviews have been conducted by expert referees, who have been requested to provide unbiased and constructive comments aimed, whenever possible, at improving the work.
3. Proceedings editors have taken all reasonable steps to ensure the quality of the materials they publish and their decision to accept or reject a paper for publication has been based only on the merits of the work and the relevance to the journal.

Title, date and place of the conference

XI International Conference on Aerospace Education and Staffing for High-Tech Enterprises (AESHE 2021)
September 21-26, 2021
Alushta, Crimea

Proceedings editor(s):

Dmitriy A. Kozorez
Vladimir T. Kalugin

Date and editor's signature

22.03.2022

Practical implementation of the algorithm for designing additional professional education programs using the competence-based approach

Rushana Anamova[1,*], *Leonid* Bykov[1], and *Dmitri* Kozorez[1]

[1]Moscow Aviation Institute (National Research University), 4, Volokolamskoe shosse, Moscow, 125993, Russia

Abstract. The paper defines the methodology for designing professional retraining programs, taking into account the requirements of professional standards and applying the "one discipline—one competence" approach as its basis. The use of the described design methodology makes it possible to formalize the designing process itself and reduce it to coordinating the requirements of the professional standard with the customer's requirements to the contents of the disciplines included in the professional retraining program.

1 Introduction

The term "additional professional education programs" refers to professional development programs and professional retraining programs [1]. The professional development programs are designed to improve the professional level of the student within a previously obtained qualification and have the duration of 16 academic hours. The contents of a professional development program are normally composed of 2 or more sections, as well as a final certification in the form of a test/exam/final assignment. The professional retraining programs are designed to prepare a student for obtaining a new qualification or carrying out a new type of professional activities and have the duration of 250 academic hours. Similarly to those of the higher education programs, the contents of the professional retraining programs are composed of several disciplines and final certification in the form of a final exam or defense of a final certification paper.

This paper is devoted to the particularities of designing professional retraining programs.

Designing the contents of a professional retraining program is a difficult and time-consuming task, since it necessitates simultaneous consideration of several sets of requirements, which are often contradictory: the requirements of the Customer (an individual or a legal entity initiating the training), the requirements of the regulatory documents (professional standards, qualification reference books, Federal State Educational Standards, etc.), as well as the actual features of the future professional activity of the students.

References [2, 3] describe an algorithm for designing professional retraining programs using a competence-based approach. The list of competencies to be obtained by a student upon completion of the training is determined based on the regulatory documents and agreed upon with the Customer. The algorithm proposed by the authors [2] implements a modular principle for building the contents of a professional retraining program, namely: a profes-

*e-mail: anamova.rushana@yandex.ru

sional retraining program is made up of modules, which themselves constitute professional development programs, and the selection of modules is carried out taking into account the competencies they form. Thus, for a professional retraining programs, the terms "discipline" and "professional development program" are identical: therefore, we will use the term "discipline" throughout the further text.

2 Methods

The competence-based approach involves building the contents of a training program, taking into account the work functions described in the corresponding professional standard. In the relevant section of the university information and analytical system, a competence base is created, which is composed of the competencies generated within the framework of studying the disciplines, meaning that each professional development program corresponds to a single competence. When designing the professional retraining programs, the necessary competencies are selected from the competence base, resulting in the generation of a list of professional development programs involved in the formation of the specific competencies. Therefore, the number of disciplines included in the structure of the professional retraining program becomes equal to the number of competencies formed during the learning process, which are selected by the methodologist and agreed upon with the Customer. Besides, it should be borne in mind that some additional competencies can be formed as a result of mastering certain disciplines within the retraining program.

Let us analyze the procedure applied to build the contents of a professional retraining program, using the ''Organization of research and development of aviation equipment'' program as an example. The trainee category is: people with higher education (specialty, master's degree).

Stage 1. Determination of the professional standard, according to the requirements of which the program contents will be formed, as well as the name of the qualification awarded (if any). In accordance with the program subject matter, the professional standard "Specialist in research and development" (approved by order of the Ministry of Labor and Social Protection of the Russian Federation No. 121n dated March 4, 2014) can be selected [4]. It should be noted that this professional standard has a code of 40.011, meaning that it belongs to the group of professional standards for cross-disciplinary professional activities in the industry.

The name of the qualification to be awarded upon completing the professional retraining program should be based on the name of the professional standard and can be formulated as follows: "Specialist in research and development of aviation equipment".

Stage 2. Determination of the list of competencies to be acquired by each student upon completion of the program. The list of competencies is formed based on the job functions (generalized job function) taken from the selected professional standard and agreed upon with the Customer.

Let us choose from the specified professional standard the generalized job function "Carrying out research and development activities within the subject matter of the organization" (code C, qualification level 6), which, in turn, includes the following job functions:

- JF-1 "Scientific management of research based on individual tasks" (code C/01.6, qualification level 6);
- JF-2 "Management of the results of research and development activities" (code C/02.6, qualification level 6).

The fulfillment of the job function JF-1 involves conducting the following job activities in accordance with the professional standard:

- JA-1 "Development of plans and methodological programs for conducting research and development on a specific topic";
- JA-2 "Arrangement of collection and examination of scientific and technical information on the topic";
- JA-3 "Analysis and theoretical generalization of scientific data in accordance with the research objectives".

The fulfillment of the job function JF-2, in its turn, involves conducting the following job activities in accordance with the professional standard:

- JA-4 "Analysis of the results of trials and observations";
- JA-5 "Implementation of the research and development results";
- JA-6 "Supervision over the correctness of the results obtained by the subordinate employees".

Therefore, for the successful fulfillment of JF-1, a student must have the following competencies (formulated based on job activities JA-1–JA-3 for the "Organization of research and development of aviation equipment" professional retraining program):

- C-1 Preparedness to develop plans and methodological programs for conducting research and development on aviation topics;
- C-2 Ability to arrange the collection and examination of scientific and technical information on the research topic;
- C-3 Preparedness to conduct analysis and theoretical generalization of scientific data in accordance with the research objectives.

For the successful fulfillment of JF-2, in its turn, a student must have the following competencies (formulated based on job activities JA-4–JA-6):

- C-4 Ability to analyze the results of trials and observations;
- C-5 Preparedness to introduce the results of research and development into production;
- C-6 Ability to monitor the correctness of the results obtained by the subordinate employees.

Stage 3. Determination of the list of disciplines forming competencies C-1–C-6. In general, a single discipline forms a single competence. There are cases when a single competence is formed by several disciplines; then, several variable professional retraining programs can be created, in which one or more disciplines differ.

The disciplines forming the specified competencies (C-1–C-6) are shown in table 1.

The discipline contents are formed based on their learning outcomes, which, in turn, are planned according to the required competence, taking into account the knowledge and skills specified in the professional standard for performing a particular job activity. For example, the learning outcomes for the "Fundamentals of Scientific Research in Aircraft Engineering" discipline can be formulated as follows:

1) K-1: know the methods of conducting research and development (wording from the professional standard for the job function JF-1);
2) S-1: be able to document the results of research and development activities (patents, scientific and technical documentation) (wording from the professional standard for the job function JF-1);
3) S-2: be able to apply the current regulatory documentation in the field of aircraft engineering (formulated based on the wording from the professional standard for the job function JF-1 "Apply current regulatory documentation in the relevant knowledge area").

Table 1. Disciplines forming the specified competencies

Competencies	Disciplines	Alternative disciplines
C-1	Fundamentals of Scientific Research in Aircraft Engineering	—
C-2	Organization of Scientific Research Activities	—
C-3	Methodology of Scientific Research	—
C-4	Methods of Mathematic Modeling	System Analysis
C-5	Fundamentals of Production Organization	Modeling of Production Processes, Production Management Systems
C-6	Quality Management Tools and Methods	Project Team Management, Project Management Strategy and Tactics

A single program section can be designed to generate one or more learning outcomes. It is also true that several sections can generate a single learning outcome. For example, the learning outcome K-1 is achieved after studying the "Methods of conducting research and development" section, the learning outcome S-1 is achieved after studying the "Patenting and protection of intellectual property" section, and the learning outcome S-2 is achieved after studying the "Industry (aviation) standards" section.

The learning outcomes and possible names of sections for other disciplines are given in table 2.

Table 2. Learning outcomes and possible names of sections for disciplines

Item	Discipline name	Competence/ Job function	Wording of the learning outcome	Name of the program chapter
1	Fundamentals of Scientific Research in Aircraft Engineering	C-1/JF-1	Know the methods of conducting research and development in the field of aircraft engineering	1.1. Methods of conducting research and development during the designing of aviation equipment
			Be able to document the results of research and development activities (patents, scientific and technical documentation)	1.2. Patenting and protection of intellectual property
			Be able to apply the current regulatory documentation in the field of aircraft engineering	1.3. Industry (aviation) standards

Table 2. (Continue)

Item	Discipline name	Competence/ Job function	Wording of the learning outcome	Name of the program chapter
2	Organization of Scientific Research Activities	C-2/JF-1	Know the means of planning research and development	2.1. Software methods and means for planning research projects
			Be able to formulate a scientific and technical problem and identify the stages of research activities	2.2. Statement of scientific research goal and objectives
			Be able to search, collect, and store scientific and technical information	2.3. Scientific search systems
				2.4. Patent search
3	Methodology of Scientific Research	C-3/JF-1	Know the methodological fundamentals of scientific knowledge	3.1. Methods of obtaining scientific knowledge
			Know the methods and particularities of theoretical research	3.2. Structure, models, and methods of theoretical research
			Know the methods and particularities of experimental research	3.3. Methodology and planning of trials
			Know the methods of processing scientific and technical information	3.4. Processing of the experimental research results
4	Methods of Mathematic Modeling	C-4/JF-2	Know the basic concepts and methods of solving computational and optimization problems using computer tools	4.1. Basic concepts and terms of mathematical modeling
				4.2. Approaches to mathematical modeling
			Be able to develop physical and mathematical models of the processes, phenomena, and objects being examined	4.3. General principles of constructing mathematical models
				4.4. Designing of various types of mathematical models
			Be able to analyze the solution obtained using computer tools	4.5. Evaluation of the results of the analysis of computational and optimization problems

Table 2. (Continue)

Item	Discipline name	Competence/ Job function	Wording of the learning outcome	Name of the program chapter
5	System Analysis	C-4/JF-2	Know the essence of the systems approach	5.1. Basic concepts of system analysis and systems approach
			Be able to apply the methods of research and development results analysis	5.2. Methods of system analysis
				5.3. Mathematical methods of decision making
6	Fundamentals of Production Organization	C-5/JF-2	Know the main production processes at the aviation industry enterprises	6.1. Characteristics of the main production processes
			Know the methods of introducing the results of research and development activities into production	6.2. Calculation of the economic efficiency of introducing the results of research and development activities into production 6.3. Methods for introducing the results of completed scientific research into production
			Know the principles of production organization and planning	6.4. Organization of production preparation 6.5. Planning of production preparation
7	Modeling of Production Processes	C-5/JF-2	Know the concept of a production system and a technological process	7.1. The concept of a production system 7.2. Technological process and production set
			Know the typical production processes	7.3. Typical production processes and their characteristics
			Know the tasks of production process optimization	7.4. Cost theory
			Be able to assess the feasibility and risks of introducing the results of research and development activities into production	7.5. Introduction of new developments into the production process flow

Table 2. (Continue)

Item	Discipline name	Competence/ Job function	Wording of the learning outcome	Name of the program chapter
8	Production Management Systems	C-5/JF-2	Know the principles of the production process organization	8.1. Organization of the production process in terms of space and time
			Know the composition and relationship of functional blocks of the enterprise production management	8.2. Information systems of enterprise management
			Know the methods of introducing the results of research and development activities	8.3. Introduction of new processes at an enterprise
9	Quality Management Tools and Methods	C-6/JF-2	Know the main stages of product quality planning	9.1. Product quality planning 9.2. Lean production
			Know the basic methods of analyzing the process failures and defects	9.3. Analysis of failure causes and consequences
			Know the process verification criteria Know the methods of verifying the results of research and development activities	9.4. Checklists
			Possess the tools for functional and cost analysis of business processes and designs	9.5. Functional and cost analysis
10	Project Team Management	C-6/JF-2	Know the methods of organizing, planning, and monitoring team work	10.1. Basic rules for managing people and small groups 10.2. Team performance assessment
			Be able to enhance the motivation of subordinates and resolve conflict situations	10.3. Motivation and conflict resolution in teamwork

Table 2. (Continue)

Item	Discipline name	Competence/ Job function	Wording of the learning outcome	Name of the program chapter
11	Project Management Strategy and Tactics	C-6/JF-2	Know the methods of making strategic and tactical decisions in project activities	11.1. Features of strategic and tactical project management
			Be able to collect and analyze information, draw conclusions about the causes and problems during the implementation of project activities	11.2. Project execution monitoring 11.3. Selection of a project strategy

The choice of disciplines forming the same competence (disciplines 4 and 5 from table 2, as well as the groups of disciplines 6–8 and 9–11), is performed upon an agreement with the training Customer.

3 Discussion and results

The disciplines presented in table 2 can be used to compile several variants of professional retraining programs with different sets of disciplines can be compiled (table 3 contains 3 examples).

Table 3. Examples of professional retraining programs with different sets of disciplines

Item	Competencies	Option 1	Option 2	Option 3
1	C-1	Fundamentals of Scientific Research in Aircraft Engineering	Fundamentals of Scientific Research in Aircraft Engineering	Fundamentals of Scientific Research in Aircraft Engineering
2	C-2	Organization of Scientific Research Activities	Organization of Scientific Research Activities	Organization of Scientific Research Activities
3	C-3	Methodology of Scientific Research	Methodology of Scientific Research	Methodology of Scientific Research
4	C-4	Methods of Mathematic Modeling	System Analysis	Methods of Mathematic Modeling
5	C-5	Fundamentals of Production Organization	Modeling of Production Processes	Production Management Systems

Table 3. (Continue)

Item	Competencies	Option 1	Option 2	Option 3
6	C-6	Quality Management Tools and Methods	Project Team Management	Project Management Strategy and Tactics

The approach applied during the designing of the professional retraining program is called a modular one. This approach allows the formation of the same set of competencies using different sets of disciplines. It should be noted that disciplines 1–3 are repeated in all three options shown in table 3; therefore, they can be combined into a single module. The advantage of the modular approach is the ability to change the set of competencies upon a Customer's request, choosing the disciplines required for the formation of the specified competencies from the list of disciplines.

4 Conclusion

Combined with the competence-based approach [5–11], the modular approach allows to design the content of an additional professional education program with minimal effort and maximum efficiency.

The algorithm for designing a professional retraining program includes the following stages:

1) determination of the professional standard, according to the requirements of which the program contents will be formed;
2) determination of the list of competencies to be acquired by a student upon completion of the program;
3) determination of the list of disciplines forming the competencies;
4) determination of the discipline contents.

An example of the variable content formation for the "Organization of research and development of aviation equipment" professional retraining program demonstrates that the application of the modular approach to designing additional professional education programs allows to reduce the program development labor costs, as well as to adapt to the Customer's requirements as much as possible.

The methodology for designing additional professional education programs, proposed in the paper, will be useful for the methodologists involved in the development of additional professional education programs, employees of training departments at enterprises, and other individuals, participating in the organization of training under additional professional education programs.

References

[1] Federal Law No. 273-FZ dated December 29, 2012 "On Education in the Russian Federation", URL `http://www.consultant.ru/document/cons_doc_LAW_140174/` (access date: 2021-10-04)

[2] R.R. Anamova, L.V. Bykov, D.A. Kozorez, Education Sciences, **10(8)**, 1 (2020)
[3] R.R. Anamova, L.V. Bykov, D.A. Kozorez, TEM J., **8(3)**, 978 (2019)
[5] Professional standard, URL https://profstandart.rosmintrud.ru (access date: 2021-10-12)
[5] N. Simone, Empir. Res. Vocat. Educ. Train., **12**, 13 (2020)
[6] C. Hofmann, K. Häfeli, X. Müller, A. Krauss, IJRVET, **8** (2021)
[7] J.J. Michael, Empir. Res. Vocat. Educ. Train., **13**, 7 (2021)
[8] I. Eegdeman, C. Klaveren, M. Meeter, Empir. Res. Vocat. Educ. Train., **12**, 9 (2020)
[9] E.L. Sorokina, Y.A. Yelinskaya, in Materials of the All-Russian scientific and practical conference Science and Society, **XII** (2019), URL https://cyberleninka.ru/article/n/osobennosti-razrabotki-i-realizatsii-programm-dopolnitelnogo-professionalnogo-obrazovaniya (access date: 2021-06-24)
[10] O.V. Fetisova, V.V. Shekhovtsov, Humanities of the South of Russia, **5** (2017), URL https://cyberleninka.ru/article/n/spetsifika-dopolnitelnogo-profes-sionalnogo-obrazovaniya-v-sovremennoy-rossii (access date: 2021-06-24)
[11] A.W. Iftikhaar, K.M. Hakim, Int. J. Humanit. Soc. Sci. Invent., **3**, 71 (2014)

Distance learning in a pandemic: the experience of sociological monitoring of students in Russia and Kazakhstan

Tanzilya Dautova[1], *Irina* Frolova[1], *Olga* Khristodulo[2,*], and *Tatiana* Rezvushkina[3]

[1]Bashkir State University, 32, Zaki Validi st., Ufa, 450076, Russia
[2]Ufa State Aviation Technical University, 12, K. Marx st., Ufa, 450008, Russia
[3]Karaganda Buketov University, Karaganda, 28, University st., 100024, Republic of Kazakhstan

Abstract. The article is analysing the experience of using distance learning technologies in higher educational institutions of Russia and Kazakhstan in the context of a coronavirus pandemic. Before the start of the pandemic, in both Russian and Kazakhstan systems of higher education, the practice of using distance learning was limited despite the digitalization trend. The challenges of supporting the continuity of the educational process have led to the fact that, in record time, teachers and students have mastered a number of platforms that ensure their online interaction. Several months of a full-scale distance learning allowed students to form a stable opinion on the new form of education. Today we have empirical material for the period of the coronavirus pandemic and we may assess the possibilities of distance learning. The authors refer to specific case studies in which students from Ufa (Russian Federation) and Karaganda (Republic of Kazakhstan) took part. In general, it can be stated that the majority of students in both Ufa and Karaganda, at the beginning of the pandemic, were not satisfied with distance learning. However, the data from 2021 revealed the opposite trend and indicated an increase in respondents' satisfaction with distance learning. The complete transition to distance education is not a pertinent question, since the potential of the traditional form of education will remain in demand. The use of proven teaching technologies will guarantee the quality and efficiency of the educational process as a whole. It doesn't matter if we're talking about digital classes or traditional ones.

In the 2000s, there were trends all over the world in the transformation of public institutions and spheres of human activity under the influence of information and communication technologies. Progress has become noticeable in the production and implementation of modern technologies, an advanced information environment has been formed, corresponding to the tasks of socio-economic development. States sought to provide all citizens with access to information resources. Preparing and ensuring the effective implementation of the transition to the digital economy also presupposed significant changes at all levels of the education system, it required expanding the information skills, increasing computer literacy, and forming critical thinking. New educational standards have emerged that secured the formation of modern competencies, including in the field of working with information, as expected learning outcomes.

In connection with the development of information and communication technologies, distance learning has become an object of increased attention among researchers. In par-

*e-mail: o-hristodulo@mail.ru

ticular, an article by Gagarina and Koldaeva [1] is devoted to innovative educational technologies of distance learning. The use of distance e-learning courses in the educational process of higher education is studied by Leontyeva and Rebrina [2], and so on. Over the course of years, the goal of creating an electronic information educational environment was set before the universities of the Russian Federation and the Republic of Kazakhstan, steps were taken in this direction on the part of both teachers and students. However, the transition to distance learning became the agenda in 2020 due to the total spread of the coronavirus pandemic. Internet resources made it possible not only to ensure the operation of educational process, but also transfered it to a more intensive and effective level. All those advantages of digital technologies that researchers have been writing about for a long time have been actualized. Thus, according to Shurygin and Krasnova, e-education enables students to develop the skills of independent work, and, consequently, to increase its effectiveness [3]. Developing this position, Shabanov asserts that e-education stimulates the student to form individual system of knowledge, skills and abilities [4]. Several researchers have pointed to the great potential of distance learning in specific academic disciplines. For instance, Seregina argued the high efficiency of distance learning in relation to mastering foreign languages [5]. A broader, comprehensive approach to the analysis of distance learning was proposed by Orlova and Koshkina [6]. Having conducted an insight into the practice of using distance education in European countries, the United States and Canada, authors analysed in detail the Russian experience before the start of the coronavirus pandemic, in which the strengths and weaknesses of this technology were tied to the financial, technical, personnel and psychological characteristics of its application. To the peculiarities of the Russian system of distance education the researchers attributed, firstly, students' habit of using printed publications, teachers relying only on their own courses, the lack of equipment for creating high-quality distance learning systems, etc. They stated that digital technologies were used as a "mediocre surrogate" of traditional education.

Indeed, the pandemic has made a significant contribution to both Russian and Kazakhstan higher education systems, as distance learning was not properly set up before. The tasks of keeping educational process operational led to the fact that teachers and students mastered a number—Zoom, BigBlueButton, Discord. The preparation of distance courses and PowerPoint presentations was also carried out intensively. There were other technical problems as well. In March 2020, when secondary school, college and university students in most countries of the world, including Russia and Kazakhstan, switched to distance learning due to the COVID-19 pandemic, almost immediately problems arose with the population's access to high-quality Internet. It was especially difficult for those who live in rural areas and remote villages to adapt to the new reality. It was necessary to take emergency measures, which, in particular, was reported by Bagdat Musin, the Minister of Digital Development, Innovation and Aerospace Industry of the Republic of Kazakhstan. From March to September 2020, Kazakhstan telecom operators invested about 60 billion tenge on a fiber-optic network in 1200 settlements of the republic. In total, according to the Ministry's Telecommunications Committee, there are 6,459 rural settlements in Kazakhstan, 4,646 of which already had the Internet in September 2020. No less ambitious processes for providing students with Internet communication resources took place in Russia.

After several months, students and teachers already had a clear opinion about the new form of education. Today we have empirical material for the period of the coronavirus pandemic and we may assess the possibilities of distance learning. To record the prevailing opinion on the issue under consideration, questionnaires of students were conducted (May 2020, June 2021) using a single methodology and tools. In total, 290 (2020) and 215 (2021) university students in the Republic of Bashkortostan were interviewed. The data obtained make it possible to establish the main trends in the understanding of distance learning by students in terms of its strengths and weaknesses.

The surveyed students were divided into the following social groups: young male—47%, young female—53%; age groups: 17–20 years old—71.4%, 21–23 years old—24.8%; 24–26 years old—3.8%; according to place of permanent residence: urban residents—65%, rural residents—35%. The survey involved students from the Bashkir State University—48.2%, Ufa State Petroleum Technological University—30.2%, Ufa State Aviation Technical University—13.9%, Bashkir State Agrarian University—3%; 4.7% of the surveyed students studied at the Bashkir State Pedagogical University and Bashkir State Medical University. According to academic "experience", the respondents in 2020 were distributed as follows: 1 course—18%, 2 course—58%, 3 course—11%, 4 course—9%, 5 course (specialty)—4% [7]. The structure of respondents in 2021 does not differ significantly from the sample of 2020 and allows to ensure the validity of the study.

The qualitative self-assessment of students in terms of the success of their studies is as follows: 14.8% of the respondents consider themselves to be "A level students", 66.8% define themselves as a "straight B student", 16% are "low-performing", and the last 2.4% are "weak students". In terms of computer technology proficiency, 17% of the respondents rated themselves as "excellent", 58.9% as "good", 21.7% as "satisfactory", and 2.4% as "poor".

The question of the advantages of distance learning is key, in our opinion. The answers to this question made it possible to build a rating line of "advantages" of distance learning. It is presented in table 1.

Table 1. "Advantages" of distance learning from the perspective of students, in %

No.	Answer options	2020	2021
1	Saves time, travel money (and money for lunch meals)	67	80
2	More time and opportunities to communicate with family	42	54
3	Adaptation of the learning time for yourself	39	52
4	Immediate clarification of problematic or incomprehensible moments via Internet resources	32	43
5	Internal discipline, self-organization rises	26	48
6	Easier to learn compared to traditional education	21	49
7	Interest and motivation to learn increases	10	27
8	The efficiency of training sessions is higher than with face-to-face classes	10	19
9	The educational material is more informative than with the traditional form of education	8	26
10	There are no "advantages"	4	0

The above data show that students see the benefits of distance learning mainly in extracurricular infrastructure components. More than two thirds (67% and 80%) of the respondents note as the main advantage "saving time and money for travel and for meals during lunchtime"; half of the respondents (42% and 54%) found that they had more time to communicate with their families. Only a quarter of respondents in 2020 and already 48% in 2021 admitted that their self-organization has increased. One in five in 2020 and half of respondents (49%) in 2021 found it easier to study. However, only 10% in 2020 and 27% of respondents in 2021 noted that their interest in learning increased and the efficiency of training sessions became higher than with traditional education (19%). It is indicative that only 8% of respondents in 2020 and 26% in 2021 admitted that the educational material in distance learning has enriched in content. It is also noteworthy that 4% of respondents (2020) do not see any "advantages" of distance learning at all.

Thus, the following conclusion suggests itself: the positive potential of distance learning in students' perception in 2020 is very modest. This can be explained, in particular, by the fact that students (and teachers), due to the rapid spread of coronavirus pandemic and hasty transition to distance learning, did not have the opportunity to thoroughly prepare for the specifics of this educational format. However, the 2021 survey showed a sharp increase in the positive potential of distance learning in the eyes of students. For almost every question, the percentage of students who answered positively in 2021 exceeded the answers to a similar question in 2020.

Nevertheless, the potential of distance learning has been revealed to an insignificant extent, and its importance in the future, in our opinion, will grow rapidly, meeting the needs of a dynamic change in the entire education system in terms of its digitalization.

It should be noted that in comparison with the traditional form of education distance learning has its drawbacks. They are ranked in descending order and are presented in table 2.

Table 2. "Disadvantages" of distance learning from the perspective of students, in %

No.	Answer options	2020	2021
1	Lack of "live" educational communication, lack of feedback	67	43
2	The load on health increases (on vision and on the spine from constant sitting in front of a computer, etc.)	64	23
3	Less opportunity to discuss issues with the teacher or with other students	59	32
4	Dependence on the speed and quality of the Internet connection (connection freeze), on the technical capabilities of the computer (or other equipment)	57	46
5	Learning material is difficult to assimilate	54	21
6	The educational material is less informative than with the traditional form of education	42	21
7	Dependence on the household noises of neighbors (repair sounds, dog barking, screaming children, loud music, etc.)	41	28
8	Some teachers have a poor command of distance teaching technology	39	36
9	This form of study does not mobilize, but discourages	39	11
10	Teachers insist on high standards less, therefore the quality of teaching suffers	21	12

As follows from the above data, in 2020 more than half of the students surveyed expressed a negative attitude towards distance learning for a very important reason—a decrease in the quality of education. More than a third (67%) of those surveyed in 2020 and 43% in 2021 recognized the lack of "live" communication and feedback in the educational process as a disadvantage.

It is important to emphasize that 59% (32% in 2021) of respondents feel the need for face-to-face discussion of educational material with a teacher and fellow students. And this is not accidental, because university education is, first of all, a live communication between a teacher and a student, as well as between the students. The concept of live dialogue is the main component of learning, through which knowledge is acquired, rethought and renewed. According to a third of the students surveyed in 2021, the online format does not reproduce social experience, which is acquired only within the walls of the university.

In 2021, almost two-thirds of respondents (64%) noted the negative health impact of distance learning. They believed that the negative impact on vision and hearing increased (since they are using headphones in both online lectures and seminars), as well as on the spine from constant sitting in front of the computer. The psychological stress from new learning technologies was also significantly high. However, adaptation to the new realities led to the fact that in 2021 the percentage of those who complained of deteriorating health dropped to 23%.

More than half of the respondents in 2020 (57%) suffered from low-quality Internet connection and obsolete computer equipment, some of them experience discomfort from the household noises of neighbors (41%). Poor sound and image quality also negatively affected the overall psychological well-being of the students. For the above reasons, more than half of the students surveyed believed that learning material is more difficult to assimilate (54% in 2020) than with the traditional form of education. By 2021, the number of dissatisfied people dropped to 21%, largely due to the improvement in the technical provision of both students and teachers. According to the respondents (39% and 36%, respectively), some teachers have a poor command of computer technologies, which additionally complicates the assimilation of educational material. In 2020, more than a third of students (39%) admited that online education demobilizes them, discourages them, and every fifth (21%) noted that teachers underestimate the level of exactingness, which affects the quality of control over material. In 2021, these indicators turned out to be significantly lower, which also indicates the effectiveness of work done in this direction.

The respondents' answers about the problems that have arisen in the implementation of distance learning, only complement the main question. Thus, 74% of respondents in 2020 and 41.7% in 2021 noted an increase in the volume of unsupervised activities. In our opinion, this is due to the very format of educational work, and the attempt of the teaching staff to play it safe against possible gaps in students competencies. According to students, in the traditional form of education, the volume of unsupervised tasks was noticeably lower. Nevertheless, students learned to cope with the growing volume of independent work, and the percentage of those concerned about this issue decreased. Among other problems, the following problems were named: health problem (55% in 2020 and 24% in 2021), mastering the material (49% and 13%), assessment of knowledge and accumulation of points (42% and 15%), technical issues (38.7% and 37%). All this also speaks of a high level of adaptation to digital educational technologies. At the same time, 8% of students in 2020 and 33% in 2021 note that "there were no special problems".

Comparing the digital indicators of answers to questions in 2020 and 2021, one can observe a sharp decrease in the number of students indicating the negative effects of distance learning. In this regard, it can be argued that during the year of study in distance learning mode, students have deeper understood its specifics and advantages and have shown a growing loyalty to the new format.

The above conclusion correlates with the students' answers to the question of their choice between online and traditional forms of education. In 2020 more than half (55.6%) of the surveyed students chose the traditional form of education, and almost every tenth (9.1%) respondent picked the distance education. In 2021 only 21.3% of respondents chose the traditional form, and 42.6% wanted to study via virtual teaching environment. Although a third of the respondents (35.3%) would prefer half of the educational process to take place in the traditional form, and half of it in the distance form.

Further, the students expressed their opinion on the effectiveness of distance learning. Almost half of the students surveyed (49% in 2020 and 44% in 2021) rated it as average, 25% of students (2020) and 8% (2021) rated it as low, 15.2% of respondents (2020) and 41% (2021) as high, and 9.7% (2020) and 5.3% (2021) found it difficult to answer this question.

Thus, in 2021, students of universities in the Republic of Bashkortostan began to rate distance learning higher.

As for the issue of the quality of education received in the distance format, only 9.4% in 2020 and 15% in 2021 considered that it is higher than with traditional education; 35.1% and 55% (respectively) of students believe that it is commensurate with traditional education, and more than half of the respondents (55.6%) in 2020 and only 19.4% in 2021 believe that the quality of distance learning training is lower than with traditional training.

Fully satisfied with distance learning in 2020 were 25% of the respondents, and in 2021—already 56% of them, partially satisfied—49% in 2020 and 36% in 2021, 20% of respondents were not satisfied in 2020 compared to 6.3% in 2021, about 6% and 2% (respectively) found it difficult to answer.

Of interest are the answers to the question, what determines the quality of education. According to students, it depends more on the professionalism of teachers than on the form of education. For example, in 2020, 31.4% of respondents agreed and 43.4% of them partially agreed with the statement that the effectiveness of the educational process mainly depends on the professionalism of teachers. At the same time, 17.2% of respondents did not agree with this statement, and the rest found it difficult to answer (8%). The students were quite critical of themselves, answering the question about the connection between effectiveness of the educational process and their personal motivation, their attitude to learning. 41% of the respondents fully agreed and 40% partially agreed with the statement that everything depends on the student's attitude to study, his or her motivation, diligence, rather than on the form of education, 12.4% of the respondents disagreed, and the rest found it difficult to answer (6.6%).

In general, it can be stated that the majority of students in 2020 were not satisfied with distance learning. However, a similar survey in 2021 revealed the opposite trend and indicated an increase in the satisfaction of distance learning among students. This phenomenon, in our opinion, can be explained by the fact that in 2020 students did not receive sufficient training in distance learning and experienced stress from a sharp and forced transition to a new learning technology. In 2021, this format turned out to be the only acceptable one in situation of total isolation. The forced transition to online learning made it possible to test new learning technologies not as an addition to traditional learning, but as the main tool capable of fully supporting the educational process.

Nevertheless, the experience of widespread introduction of distance learning into the educational process is still insignificant. The existing technical, methodological, communication and psychological barriers, which currently prevent teachers and students from improving the effectiveness of distance learning, are not of a fundamental nature. In our opinion, having eliminated the identified shortcomings, this form of education could become not only an important addition to traditional education, but it would be able to act as an independent form of education in some disciplines.

As for the students themselves, they identified as the priority areas for improving distance learning the following:

- facilitation of teacher–student and student–student communication (47%);
- provision of learning options (lecture recordings, basic, additional and reference literature, assignments for unsupervised work, etc.) (46%);
- elimination of technical interference (39%);
- adaptation of the teaching methodology for distance classes (37%);
- improvement in qualification and competence of teachers in the use of distance technologies (31%);
- ability to attend lectures by prominent lecturers from other universities inside the country and worldwide (28%) [2].

We now turn to the assessment of distance learning by students of Karaganda universities. We would consider the data from a study, in which the transition to distance learning was analysed. As part of the project, implemented in March–May 2021, interviews and focus groups were conducted, in which more than 110 students from Karaganda universities took part. Here's an example from an interview with one of them.

Edil is 19-year-old, and he is currently a third-year sociology student at the Karaganda University. His tuition is free, as he is studying on a state educational grant. In order to attend lectures, participate in seminars and keep up with the educational program, Edil bought a laptop on credit and returned home—to the village of Zhairem, Karaganda region, since living in Karaganda is expensive. He was told that the Internet was installed in his home village. In fact, it turned out that the speed of the Internet does not allow to fully engage in online activities:

> Connection freezes, sometimes I can't connect, because of this I lag behind, teachers complain about me. Even when you watch a record, the speed is still not enough [...] Since the beginning of the online [classes], I have already connected mobile internet several times, as [the local Internet] does not really work. It takes a lot of money.

The introduction of information technology into the educational process does not make sense if electricity is cut off in rural areas. This is exactly what happens in Edil's village, where electricity is often cut off. Although, how you can listen to lectures using a slow mobile connection is also a big question.

> Last year I was unable to pass the entire session. Nowadays the light is not turned off so often, but before that it was off three or four times a week. Therefore, I warn teachers in advance. And when I send out my home assignments, I also worry that the light might be turned off.

Edil is worried that he was considered lately as a "low scorer". Teachers are unhappy with his knowledge, although they understand that in rural areas there are problems with access to a high-quality Internet connection:

> I may need to sit in Zoom classes for five hours. This program requires a good Internet, but it comes and goes. If the Internet is bad, then the Zoom does not load, every time I wait for 10–15 minutes and then I come late to classes. Loading tasks, it is generally better [to do that] after 12 at night, when the Internet works normally. Only after midnight I can download all the tasks.

Edil wants to return to the old format of education, and not only because of the poor Internet connection. He also complains about health problems, when he has to sit in front of the computer for hours, which causes problems with his back and vision.

> I have vision problems. Now, due to the fact that I sit in front of the computer much longer than before I've noticed that my vision is deteriorating, although a little time has passed. Studying is from 8 am to 5 pm, and I study almost without getting up, I don't even have time to dine. I want to study offline. When you sit in front of the teacher, you understand everything. So far I cannot say that I am fully acquiring knowledge.

The situation described is typical for students living in rural areas. Lucky considered those students who have relatives in the city that allow them to temporarily stay and study properly. For example, Dilyara has been living in Karaganda with relatives since September. She comes from one of the small towns of the Karaganda region, which is located 40 kilometers from the city:

> This is a temporary measure ... I won't stay with [relatives] forever ... If there was Internet at home, I would of course stay at home and study online. I get out of the situation by living with my aunt ... We've been promised to have Internet in our house in the fall. Last year the money was collected and the Internet was supposed to be installed, but it didn't happen. Now we look forward to it this year.

As a positive side of the transition to online education, students of universities in Kazakhstan noted the possibility of combining studies with work. Working in the service economy, in the positions of waiters, salesmen, consultants, managers, students have the opportunity to be employed and progress in studies instead of taking an academic leave or spoiling their reputation as "good students" by skipping classes and constantly asking for time off in order to go to work. In addition, they retain the ability to independently pay for their education or make a certain contribution to the family budget. Here is a typical opinion of a working student:

> It is convenient when you need to combine work with personal affairs. For example, I work from morning to evening. During online classes I just go to the break room, do the tasks. If the classes were offline, then I would have to leave work, spend time on the road, ask the superiors for time off and explain where I am going.

The interviewed students said that over the past year they got used to the distance learning format and adapted to its shortcomings:

> I feel good about it, since we have been sitting at home for 1.5 years already, the lectures are in Zoom, and everyone is already used to it. We have already begun to understand how everything needs to be done.

When assessing the quality of teaching with distance learning, students said that most of the teachers quickly mastered the new software, although there were those who could not adapt to the online format:

> The teachers got ready very quickly and got used to it ... From the first classes we tried various platforms, where it was possible to work in pairs.

According to the results of the research carried out in Karaganda, it can be stated that the assessment of distance education from the point of view of students from Kazakhstan is ambiguous. On the one hand, students complain about the instability and high cost of the Internet, on the other hand, they like the fact that it is possible to save money on travel and combine study with work. Their comments on the future of distance education are also ambivalent: some believe that it is necessary to preserve the traditional format of education, others—that it should be a blended learning, and still others find that distance learning should prevail.

Thus, analysing the data from sociological monitoring of university students in the city of Ufa (Russian Federation) and Karaganda (Republic of Kazakhstan), we can conclude that distance learning has great potential, the implementation of which will significantly increase the efficiency of the educational process in general, especially in the system of higher education. However, the question of a complete transition to distance education is premature, if not inappropriate. The potential of the traditional form of education will remain in demand due to the unique opportunities that distance education cannot provide. The use of proven teaching technologies will guarantee the quality and efficiency of the educational process as a whole. It doesn't matter if we're talking about digital classes or traditional ones.

The authors of the article are thankful to the Science Committee of the Ministry of Education and Science of the Republic of Kazakhstan for the research materials, which were obtained in the scientific project AR 8052518 "Formation of information culture of Kazakhstan's youth as a factor in the growth of the intellectual potential of society (2020–2022)"

References

[1] P.A. Gagarina, V.D. Koldaev, *Innovative Educational Technologies of Distance Learning*, Economic and Social Research, **4**, 96 (2018)

[2] I.A. Leontyeva, F.G. Rebrina, *Application of Distance E-Learning Courses on the Educational Process of Higher Education*, Bulletin of Chelyabinsk State Pedagogical University, **3**, 114 (2018)

[3] V.Y. Shurygin, L.A. Krasnova, *Electronic Learning Courses as A Means to Activate Students' Independent Work in Studying Physics*, Int. J. Environ. Sci. Educ., **11(8)**, 1743 (2016)

[4] A.G. Shabanov, *Distance Learning in the Context of Lifelong Education: Problems and Development Prospects* (SSU Publishing House, Moscow, 2009)

[5] E.A. Seregina, *Distance Learning Technology as a Way of Getting Higher Education in Russia*, Perspectives of Science and Education, **2**, 25 (2018)

[6] E.R. Orlova, E.N. Koshkina, *Distance Learning: Realities and Prospects*, Bulletin of the International Institute of Economics and Law, **1**, 5 (2011)

[7] T.A. Dautova, *Distance Learning Through the Eyes of Students (on the Example of the Republic of Bashkortostan)*, Economics and Management: Research and Practice Journal, **4**, 146 (2020)

https://doi.org/10.1051/shsconf/202213701003

Problems of the university teaching transition to e-learning

Ilya Ginzburg[1], *Anastasia* Denisova[1], *Alexandr* Ermakov[1,*], and *Nikita* Shurygin[1]

[1]Moscow Aviation Institute (National Research University), 4, Volokolamskoe shosse, Moscow, 125993, Russia

Abstract. The article discusses the higher educational institutions transition problems to a distance learning operation mode during the pandemic, taking into account the technical universities work specifics. The article describes the problems that have arisen as a result of e-learning mode transition, as well as some solutions to these problems. The article focuses on: the teachers' readiness problems; the e-courses quality for technical specialities students; the teachers' developed materials copyright; virtual learning environments to replace the laboratory stands and equipment for students in distance learning process; distance learning educational process implementation technical support problems. The described problems' proposed solutions' implementation can be carried out in higher educational institutions without hiring additional personnel.

1 Introduction

The situation in higher education institutions during the coronavirus COVID-19 pandemic period highlighted various problems concerning the institutions work organization, the relationship between personnel, teachers and students, financing the development of higher educational institutions and their material and technical base equipment [1–3].

But despite the problems and difficulties that have arisen, thanks to the distance learning information technologies introduction, it was possible to ensure the distance learning educational activities implementation, to achieve students' high qualifications and appropriate levels of professional competencies' formation [4].

A general problem solving approach for uncertainty conditions was developed to let the new solutions faster introduction. Acceleration became possible due to an iterative process, which made it possible to start working on solving problems without initially having the complete information and obtaining the necessary data in the course of work.

2 Technical problems

First, it is worth mentioning the technical problems that have arisen as a result of a transition to a total remote operation mode that students and teachers have encountered. Among such technical problems the most frequently encountered:

*e-mail: ermakovaa@mai.ru

2.1 Individual computer facilities insufficient provision

Individual computer facilities insufficient provision (including personal computers (PC), mobile devices like laptops and smartphones) has become the reason for the educational activities' mode desynchronization in remote operation mode.

To solve this problem, it is required to equip teachers with PCs or laptops with a camera and a microphone to meet the distance learning management systems (LMS) requirements.

Another difficulty in determining the presence and severity of this problem lies in the fact that even the provision of users with personal computing means does not mean automatic compliance of the available computing facilities with the system requirements for software and hardware of the implemented LMS, communication and collaboration platforms (CCP) and other programs used in the distance learning process. In this regard, one should very carefully choose the software to be introduced so that as many of the available computing devices as possible could be used to work with the introduced software. Today, the global trend is the use of web applications and virtual machines' remote access tools wherever possible. This allows the hardware and software requirements to be minimized to: a stationary or mobile device with a standard web browser. With the widespread adoption of HTML5-capable web browsers, large and complex systems can be built as web applications. During the pandemic, the more intensive use of web applications and remote access server infrastructure by a large number of users has demonstrated some performance and bandwidth problems.

2.2 Server infrastructure insufficient performance

The insufficient server infrastructure performance becomes notable with a sharp increase in the number of LMS users and the transition of technical students' laboratory work from test benches based on real laboratory devices to virtual environments with multi-user access through communication channels. Before the pandemic, the server capacities of many educational institutions were not designed for such a load, and the performance of the server infrastructure had to be rapidly increased.

To scale server infrastructure fast one can use the cloud computing systems for virtual servers and various applications deployment. This solution is the fastest, easiest to use, highly scalable and is often cheaper than building your own infrastructure of similar performance. The other way is to use the third-party data centers for hosting own servers or renting their servers for the server infrastructure scaling. Next goes establishing of own high-performance virtualization servers, which can be quickly incorporated into the existing infrastructure to transfer existing virtual machines to the powerful virtualization servers with minimal reconfiguration. And finally there comes creation of scalable multi-server information systems that can operate on server clusters with different computing performance, ensuring performance scalability. This multi-server approach allows combining external cloud servers, physical servers and virtual machines in one cluster.

The server infrastructure development strategy, containing a reasonable combination of the proposed solutions, will make the server infrastructure of an educational institution more versatile, suitable for solving a changing range of tasks and easily scalable.

The optimal combination of solutions depends on for what tasks, how quickly, for how long, and with what budget, you need to create a server infrastructure.

2.3 Communication channels' insufficient bandwidth

The insufficient communication channels' bandwidth becomes noticeable with an increase in the number of simultaneously connected users to the server infrastructure and an increase in the traffic volume that users exchange with the infrastructure. The bottleneck is usually the channel through which the server infrastructure is connected to the Internet. Users working remotely from home usually use individual communication channels that are designed for much bigger amount of traffic than is required for work. Problems arise with a server infrastructure channel that was not originally designed for such a load.

To solve this problem one can use the cloud services and content distribution networks designed to serve for many users and high traffic or use third-party data centers to provide reliable user access to own servers or develop own network infrastructure, which gives a new impetus to the development of the organization's information systems, but can be a very expensive undertaking, and therefore requires careful planning.

The network infrastructure development strategy, containing the required combination of the proposed solutions, will make the network infrastructure of an educational institution more versatile, suitable for solving a changing range of tasks and easily scalable.

The optimal combination of solutions depends on what characteristics, how quickly, for how long, and with what budget, you need to create a network infrastructure.

2.4 Live communication with the teacher process violation

Sometimes questions or situations arise that are impossible to discuss or analyse in distance learning without personal communication.

To solve this problem, in addition to web LMS, such as LMS Moodle, the CCPs are used, such as Microsoft Teams, Zoom, Discord, BigBlueButton, etc. This makes it possible to ensure all participants' live presence at the lesson in one virtual classroom "with a whiteboard and a projector". Critics of the CCP use say that some students lose motivation as opposed to being physically present in the classroom. Some of the observations made have shown that those students who are not motivated to study in the CCP usually do not have the motivation to study in the classroom, so there is most likely no any significant difference with physical presence in the classroom for university students. But when studying in the CCP, such students do not demotivate and do not distract others during the lesson. And if someone has problems with the material perception or with the learning motivation, then students can always discuss these problems with the teacher in the same CCP. As for the argument about the teacher positive personal impact with physical presence on those students who, as we said, lack motivation, this impact often boils down to comments about the distracting behavior of these students in class. Some measurements of academic performance do not show a significant positive effect of physical presence alone on the low-motivated students' performance. However, we can conclude that being distracted to increase the low-motivated students' motivation during classes the teacher obviously spends less time on directly explaining the new teaching material for all students. This cannot positively affect their understanding of the material, reduces the amount of material discussed in the class, and may also reduce overall academic performance. To increase motivation and discuss difficult issues, teachers can use the CCP during their office hours to attract more students, who no longer need to come to the teacher's office.

2.5 Discrepancy in time depending on the time zone

Discrepancy in time depending on the time zone does not allow starting distant classes at the same time (if e.g. a student studies in GMT+1 time zone, but lives in GMT+9).

To solve this problem, you should record the lessons in the screencast form, and then upload the media files to the web LMS to let the students in different time zones watch the lessons. The students' questions should be discussed on the web LMS forum to let the discussion be available to all class participants.

If there is a slight discrepancy in time, the schedule can be changed so that classes in the CCP are held for all students at work time (or at least not at night).

2.6 Difficulties associated with the knowledge control and certification processes

Some difficulties arise associated with the students' knowledge control and certification processes [5, 6] on distance learning.

To solve these accounting problems for the students mastering the material, one should use electronic sheets, e.g. provided by the LMS Moodle, which can be configured to account for progress in various assignment types with different grade types. But the distance mode with LMS and written tasks does not guarantee the knowledge quality demonstrated by students.

This problem is successfully solved by using the CCP for live communication with the student's camera and microphone turned on, which allows the needed interaction. The only difficulty may remain is the time zone difference, but you can choose a time that suits everyone, or hold several online meetings for different time zones.

For credits, examinations and final qualifying work defence, the CCP should also be used with the student's webcam and microphone turned on, including when a student is simultaneously connected from a stationary (computer or laptop) and mobile (smartphone or tablet) client device to use multiple microphones and webcams for proctoring.

3 Organizational and technical problems

Among the more serious problems that cannot be solved by technical means alone, the following were noticed:

3.1 Copyright protection for digital materials placed in the LMS

Copyright protection for digital materials developed by the teachers and placed in the LMS to provide distance learning [7] is a concern source for many teachers.

The solution to this problem should include tracking and suppressing the digital materials authored by university employees dissemination by unauthorized persons and should be a comprehensive and long-term activity.

For some dynamically changing training courses (e.g. on the modern technologies), it may turn out that teaching materials regularly change significantly enough that even their possible distribution outside the LMS does not pose a serious threat to this course attractiveness for study at the university, but, on the contrary, could act as an advertisement.

Also, for advertising purposes, short open courses can be created, e.g. for schoolchildren or for a wide range of people without special training.

3.2 A decrease in the research activities productivity

The pandemic caused a certain decrease in the research activities productivity, especially when it comes to a technical or applied profile (sculpture, painting, etc.) university [8]. For all activities requiring access to various equipment that cannot be located at home [9, 10] the transition to remote mode is the reason for the work complete cessation.

To solve this problem, it is required to use telepresence tools (including remotely controlled avatar-robots), as well as laboratory equipment control consoles and sensor data output to web interfaces. This will ensure that experiments are carried out with a minimum personnel physical presence and the experiment participants able to connect through public communication channels using secure virtual private network (VPN) connections.

3.3 A Decrease in the students' skills and competencies formation quality

The pandemic affected a certain decrease in the future professionals' skills and competencies formation quality in the technical field, where classes are often held using special technical training aids, installations, laboratory equipment, etc. The situation is aggravated when it comes to practices that must be carried out in conditions close to production activities [11–13]. The most serious problem of a complete transition to a remote operation mode has affected technical higher educational institutions.

The solution to this problem is partly similar to the solution to the previous one, but it is also possible, taking into account the laboratory educational installations functioning [14, 15], to create their virtual simulators using the virtual desktop infrastructure (VDI) technology or a web interface, available to students at any time via the Internet, which need much less maintenance than physical devices. The students' work results with this equipment can be registered directly in the LMS and sent for verification to the teacher.

4 Organizational problems

Speaking about the technical specialities students' teaching difficulties in a pandemic that have caused a complete change in the technical university educational process and it's organization technology, it is necessary to note the students' practical training approaches severe change extreme necessity. This entailed the need to comprehend a number of organizational problems, including:

4.1 Fully developed electronic courses lack or absence

Fully developed electronic courses lack or absence, taking into account the practical and laboratory work implementation, which are performed using special production equipment (e.g. laboratory stands, machine tools, visual aids in mechanics, hydrostatics, kinematics, etc.), as well as the lack of time in the teachers' working schedule to create text and media materials specifically for LMS has become a problem.

To solve this problem, it is necessary to allocate additional working time for teachers [16] to create digital content for LMS, as well as train university teachers in effective practices for creating digital content.

The students can also help the teachers to create the test questions' banks for LMS. First, the questions are formulated by the teacher. At this stage various similar questions should be formulated to make cheating difficult. Then sets of questions are entered into the LMS and offered to students as midterm control. In the first iteration, students should formulate short answers to these questions themselves. After checking, the teacher receives many correct and incorrect answers, which can be combined with the developed questions to create a test questions' bank with variable options for correct and incorrect answers.

4.2 Virtual learning environments lack or absence

Lack or absence of virtual learning environments implemented in a technical university educational process, which allow distance learning students to adequately replace the laboratory equipment use (e.g. device simulators, etc.) has also become a problem.

To solve this problem, it is necessary to allocate resources to create virtual learning environments, as well as to attract teachers in disciplines where it is planned to use such environments to this development with the additional working time allocation for teachers.

4.3 A Severe teaching approaches' change

A severe study regime change during a pandemic caused a severe teaching approaches' change (e.g. when studying complex topics of physics, chemistry and other disciplines and educational modules), which required the new didactic-methodological tools rapid development. E.g. in chemistry complex topics teaching process, teachers had to replace classroom lessons using visual experiences and experiments in which students themselves took part, with video materials in which it is difficult to convey all the experiment nuances, answer all students' questions and analyse the difficulties that arise. This caused certain difficulties in the students' knowledge, abilities, skills and competencies formation, as well as in their verification within the current and intermediate certification framework.

It has now become clear that distance learning using information technology is an educational system new everyday version, so it is necessary to create some new didactic and methodological tools, taking into account their use in distance learning. This will ensure the proper quality of education in the new environment.

4.4 Lack of proper technical support for the distance learning implementation

Lack of proper technical support for the distance learning implementation in the technical universities' educational process in many specialities and areas where the distance learning should be used has made the distance learning transition a slower process. With the massive transition to distance learning, the task was to quickly train university staff in distance learning technologies. This has become a problem in many specialities since approaches to their distance teaching were not developed in advance.

In the future, this problem can be solved by involving the practitioner-specialists in certain disciplines together with those who are already conducting students' training in these disciplines in the distance learning and teaching methods development.

4.5 Lack of trained teaching staff for remote teaching

Switching to distance learning has shown the lack of specially trained qualified personnel, who could effectively carry out students' remote training, especially for those disciplines' teachers where student training takes place using training stands, equipment, apparatus, etc.

A solution to this problem can be achieved within the advanced university teachers' training courses framework, where the university teachers will be trained in the distance work effective organization using various means of LMS, CCP and ways of distance learning digital content creation.

4.6 Structuring teaching material complexity in a video lecture

Structuring teaching material in a video lecture requires a large accompanying material amount, such as graphs, diagrams, formulas, calculations, etc., which the teacher needs to use simultaneously or constantly refer to them during the lesson.

This problem is successfully solved by preliminary opening several application windows demonstrating the necessary materials, expanding these windows to the teacher's PC full screen, and then organizing the teacher's screen display to the students through the CCP with switching between open windows with educational materials. This allows using various materials and software installed on the teacher's PC during the lesson, and quickly switching between them.

4.7 Insufficient distance learning quality in technical/applied disciplines

Whatever the educational material educational potential in the LMS, live contact with the teacher is necessary for technical/applied disciplines university students.

This problem relates primarily to text and graphic materials and is solved by adding media materials to the LMS and using the CCP for live communication.

5 Conclusion

At present, the complete replacement of many academic disciplines/educational modules teaching with distance learning does not allow ensuring the proper education quality [17]. However, in those disciplines for which the problems listed in the article have been solved, it is now possible to ensure the education quality similar to face-to-face education and even surpassing it by facilitating the students' independent work with digital educational materials in the LMS and increasing the teachers' contacts availability with the CCP. The distance learning benefit is very high in organizing the individual learning trajectory and the self-study process for the students in a convenient mode. Distance learning allows the teacher to implement group and individual support for the students.

The proposed solutions to the described problems could be implemented in higher educational institutions without hiring additional personnel. That eases the adoption of the e-learning education process [18, 19].

As a result, we can talk about the full-fledged blended learning emergence in which distance e-learning complements the traditional face-to-face education and is already an integral part of a technical university educational activity.

References

[1] T.A. Kamarova, *Signs of precarization in a pandemic*, Professional education and labour market, **2(41)**, 74 (2020), DOI: 10.24411/2307-4264-2020-10222

[2] P.N. Osipov, *Viral digitalization and its consequences*, Professional education and labour market, **2(41)**, 75 (2020), DOI: 10.24411/2307-4264-2020-10225

[3] N.N. Skrypnikova, *The education future: total distance or total rejection of it?*, Professional education and labour market, **2(41)**, 58 (2020), DOI: 10.24411/2307-4264-2020-10213

[4] L.V. Sidorova, Yu.V. Krupskaya, *Pros, cons and online education prospects*, Professional education and labour market, **2(41)**, 87 (2020), DOI: 10.24411/2307-4264-2020-10229

[5] O.V. Ermakova, M.N. Kaloshina, E.V. Dianova, *Management of Innovative Projects over the Life Cycle of Distributed Aviation Systems*, Russian Engineering Research, **39(5)**, 439 (2019)

[6] O.V. Ermakova, N.V. Ruban, *Improving the terms of the life cycle contract for the creation of complex technical objects in high-tech industries*, J. Phys.: Conf. Ser., **1515**, 032017 (2020), DOI: 10.1088/1742-6596/1515/3/032017

[7] I. Vlasov, *Legal and pedagogical aspects of e-education*, in International Conference Engineering Technologies and Computer Science (EnT), 24–26 June 2020, Moscow, Russia, 144 (2020), DOI: 10.1109/EnT48576.2020.00034

[8] L.I. Yashina, O.M. Goreva, *Distance education implementation problems in the university*, The Surgut State Pedagogical University Bulletin, **4(61)**, 84 (2019)

[9] S. Borisov, Y. Glukhovskaya, S. Dobrovolskiy, P. Nikitin, I. Podporin, *Concept of an Experimental Setup for Testing the Technology for the New Anti-Corrosion Coating Materials Formation Using Low-Temperature Supersonic Heterogeneous Flows*, TEM J., **9**, 566 (2020), DOI: 10.18421/TEM92-19

[10] V.A. Zagovorchev, O.V. Tushavina, *The use of jet penetrators for movement in the lunar soil*, INCAS Bulletin, **11**, 121 (2019), DOI: 10.13111/2066-8201.2019.11.S.22

[11] Y. Kondrashov, O. Glushkova, D. Kobzev, *Planning and Approving Corporate Resource Development*, in Silhavy R., Silhavy P., Prokopova Z. (eds.), Software Engineering Perspectives in Intelligent Systems: Proceedings of the Computational Methods in Systems and Software, **1294** (Springer, Cham, 2020), DOI: 10.1007/978-3-030-63322-6_86

[12] A. Stankevich, *A Model for the Operating Management of the Aircraft Maintenance Composition*, in R. Silhavy, P. Silhavy, Z. Prokopova (eds.), Software Engineering Perspectives in Intelligent Systems: Proceedings of the Computational Methods in Systems and Software, **1294** (Springer, Cham, 2020), DOI: 10.1007/978-3-030-63322-6_89

[13] O.V. Ermakova, *Assessing the Difficulty and Cost of Research and Development for the Case of Distributed Aviation Systems*, Russian Engineering Research, **38(6)**, 462 (2018)

[14] V.A. Zagovorchev, V.V. Rodchenkov, *Calculation of Main Design Parameters of a Reactive Penetrator Intended for Movement in Lunar Soil*, Russian Aeronautics, **62(4)**, 595 (2019), DOI: 10.3103/S1068799819040093

[15] V.A. Zagovorchev, O.V. Tushavina, *Selection of Temperature and Power Parameters for Multi-Modular Lunar Jet Penetrator*, INCAS Bulletin, **11**, 231 (2019), DOI: 10.13111/2066-8201.2019.11.S.23

[16] A.S. Zinchenko, *Project-focused personnel management approach of higher educational institutions*, Asia Life Sciences, **Supp22(2)**, 243 (2020)

[17] Mulyani, Fidyati, Suryani, M. Suri, Halimatussakdiah, *University students' perceptions through e-learning implementation during COVID-19 pandemic: Positive or negative features dominate?*, Studies in English Language and Education, **8(1)**, 197 (2021), DOI: 10.24815/siele.v8i1.17628

[18] I.G. Bakanova, E.A. Yelizarova, L.V. Kapustina, *Significance of the Organization of E-Learning Management System in a Modern University*, in S.I. Ashmarina, J. Horák, J. Vrbka, P. Šuleř (eds.), Economic Systems in the New Era: Stable Systems in an Unstable World. IES 2020. Lecture Notes in Networks and Systems, **160**, 467 (Springer, Cham, 2021), DOI: 10.1007/978-3-030-60929-0_60

[19] Q. AlAjmi, R.A. Arshah, A. Kamaludin, M.A. Al-Sharafi, *Developing an Instrument for Cloud-Based E-Learning Adoption: Higher Education Institutions Perspective*, in S.K. Bhatia, S. Tiwari, S. Ruidan, M.C. Trivedi, K.K. Mishra (eds.), Advances in Computer, Communication and Computational Sciences. Advances in Intelligent Systems and Computing, **1158**, 671 (Springer, Singapore, 2021), DOI: 10.1007/978-981-15-4409-5_60

Education of students for the staffing of high-tech industries at the GDD Department at Novosibirsk State Technical University

Anatoly Guskov[1], *Konstantin* Milevsky[1], *Igor* Balagansky[1], and *Alexey* Vinogradov[1,*]

[1]Novosibirsk State Technical University, 20, Prospekt K. Marksa, Novosibirsk, 630073, Russia

Abstract. The paper deals with the aspects of staff education for the high-tech industries of Russia at the GDD Department at Novosibirsk State Technical University. The peculiarity of training is the development and implementation of targeted educational modules, taking into account the peculiarities of the enterprises where the graduates will work. The annual All-Russian student conference is held to develop the skills of scientific activities, public presentation, and approbation of the results. In the process of training and holding the conference the resources and employees of the institutes of the Siberian Branch of the Russian Academy of Sciences are involved.

1 Introduction

The Department of Gas Dynamic Impulse Devices (GDD) was created in 1985, by the decision of the Ministry of Higher Education No. 701 from 4.11.1985 [1].

The department concentrated on the training of mechanical engineers in the specialty 0546 "Production of shells". At present this specialty is called 170103 "Means of Defeat and Ammunition".

This specialty produces highly qualified engineers with a wide profile, specializing in the design and production of a wide range of weapons and ammunition, methods of processing, and creation of materials and products with unique properties. Students receive in-depth knowledge in a complex of basic engineering disciplines, explosion physics, deformable solid mechanics, internal and external ballistics, design, and technology of ammunition production.

Graduates are trained to develop modern high-energy devices and technologies for using explosion energy in various industries:

- in mechanical engineering—for shaping, hardening, welding, and cutting materials;
- in materials science—for the synthesis of diamonds, cubic boron nitride, and high-temperature superconductors;
- in mining—for exploration and extraction of minerals;
- in construction—for moving large volumes of rocks, for drilling works, for the destruction of old structures and etc.

*e-mail: vinogradov@corp.nstu.ru

2 Targeted education

Each year the number of applicants for the special-purpose places by orders of high-tech industries of the city of Novosibirsk and the Novosibirsk region increases, and the number of the enterprises that send school graduates for training at Novosibirsk State Technical University also expands. At present, the Department of GDD carries out point annual special training of about 10–12 students for 5 enterprises of Novosibirsk region: JSC Institute of Applied Physics, JSC NMZ Iskra, Kuibyshev Chemical Plant, FKP Novosibirsk pilot plant of measuring instruments, Novosibirsk plant of artificial fiber, Novosibirsk cartridge plant.

Within the target program "New Personnel for the High-Tech Industries", since 2015 the department has been training the following educational modules:

"Training of highly qualified specialists in the field of development of firing means" for the enterprise JSC "Institute of Applied Physics". The developed courses "Fundamentals of Designing Engines for Dischargeable Means", "Designing Weapons, Ammunition, and Dischargeable Means", "Industrial (pre-diploma) Practice: Practice to Obtain Professional Skills and Experience", "Ammunition and Dischargeable Means Design" of the educational module are taught not only at the department but also on the premises of the enterprise.

The educational module developed together with the employees of IPF JSC takes into account all the requirements for the competencies of students in the "Ammunition" direction, which are laid down in the design of new, modernized and advanced firing means.

Students who passed the training and attestation on the educational module are ready for the production, design, and technological and research professional activities in the design and technological preparation of the production of ammunition as a designer, research engineer, and developer.

The material and technical base of JSC NMZ Iskra was used to train highly qualified specialists in the field of explosive devices development for some disciplines of the educational module: research work, internships, and graduate qualification works.

Due to the customer's new requirements for the efficiency and economy of blasting equipment and the expansion and modification of the product range, the company needs qualified personnel capable of designing, developing production processes, and testing modern blasting equipment for various purposes and adapted to the company's possibilities. In this regard, the developed educational module will be in demand and will have a significant scientific and production significance.

Students who have passed training and certification on the educational module are ready for industrial, design, and technological and scientific research professional activities in the design and technological preparation of the production of explosive devices as a designer, research engineer, and developer of technological production processes.

The educational module "Development, production, testing and utilization of cartridges for small arms and small-caliber artillery weapons" is developed in cooperation with JSC "NPZ". Students of the department annually undergo practical training, where they are given the opportunity to get acquainted with the production, technological process of cartridges, and cartridge cases production.

Students who have passed training and certification on the educational module have competencies in designing and developing technological processes for the production of shells and cartridges.

Graduates of the department upon graduation are employed as design engineers, as well as process engineers, where they gain experience in cooperation with experienced engineers at the plant, and subsequently implement their skills becoming leading specialists.

The educational program of retraining of personnel—designers, researchers, testers, technologists for the defense industry in the field of the life cycle of manufactured products is conducted.

3 Conferences

Within the framework of approbation of the results students of the target, direction takes part in all-Russian and international conferences: "Science. Industry. Defense", "Science. Technologies. Innovations" Novosibirsk, "Reshetnev readings" Krasnoyarsk, "Actual problems of modern continuum mechanics and celestial mechanics" Tomsk.

All-Russian Conference of Students, Young Scientists and Postgraduate Students "Science. Industry. Defense" has been held annually since 2000 [2]. The conference was initiated by the NSTU Department of Gas Dynamic Impulse Devices, which trains specialists for the high-tech industries. Since 2001, students and graduate students from other universities in different Russian cities started to participate in the conference: Novosibirsk, Vladimir, Moscow, Saint Petersburg, Nizhny Tagil, Novokuznetsk, Samara, Tomsk, and others. The aim of the conference is to intensify scientific research, establish contacts between institutes of higher education in the country, create common information and scientific space, instill in students an interest in scientific research. Representatives of industry and interested organizations have the opportunity to meet and discuss their problems with scientists-specialists, make a presentation and participate in the discussion.

The conference was also attended by scientists from institutes of the Siberian Branch of the Russian Academy of Sciences (the Sobolev Institute of Mathematics, the G.K. Boreskov Institute of Catalysis, the A.V. Nikolaev Institute of Inorganic Chemistry, S.A. Khristianovich Institute of Theoretical and Applied Mechanics, the S.S. Kutateladze Institute of Thermophysics, Russian Academy of Sciences, Institute of Solid State Chemistry and Mechanochemistry, Institute of Systematics and Ecology of Animals, Institute of Petroleum Geology and Geophysics, Institute of Computational Mathematics and Mathematical Geophysics), staff of scientific-industrial enterprises (S.A. Chaplygin Siberian Research Institute of Aviation, V.P. Chkalov Novosibirsk Aircraft Plant, Nanostructured Coatings Ltd, Scientific and research Research Institute of Electronic Devices, M.F. Reshetnev Information Satellite Systems).

The conference is divided into 14 sections: design and operation of weapons and munitions; high-energy condensed systems; industrial processes; rocket and space technology; aircraft control systems, dynamics and strength of machines, aircraft and helicopter construction and operation of aircraft and engines, technique and physics of low temperatures; aircraft life support and protection systems; hydroaerodynamics; safety of technological processes and industries; ecology; environmental management; environmental protection; economics and management in the industry.

Each conference is held with a patriotic focus in honor of a grand event in the historical calendar of the Fatherland or a person who has left an indelible memory in the minds of his contemporaries. For example, in 2001 it is the thirtieth anniversary of "Means of Defeat and Ammunition" specialty, in 2002—the memory of A.I. Belosokhov, graduate of "Means of Defeat and Ammunition" specialty, in 1996—appointed the first deputy minister of the Russian Federation on atomic energy, in 2003—110th anniversary of Novosibirsk, in 2005—60th anniversary of the victory in the Great Patriotic War, in 2006—90th anniversary of the professor, doctor of technical sciences, Honored Worker of Science and Technology, Rear Admiral G.S. Migirenko, in 2007—the 50th anniversary of the Siberian Branch of the Russian Academy of Sciences, in 2008—115th anniversary of Novosibirsk, in 2009—50th anniversary of the Faculty of Aircraft NSTU, in 2010—65 Anniversary of Victory in Great Patriotic War, in 2011—50th anniversary of the first space flight, in 2012—200th anniversary of the Battle of Borodino, in 2013—100th anniversary of Novosibirsk pilot, thrice Hero of the Soviet Union, Air Marshal A.I. Pokrysishkin. In 2015—the 70th anniversary of Victory in the Great Patriotic War.

The conference consists of three days: first day—registration of participants and their resettlement. Plenary session, where prominent scientists speak.

Day 2—speeches of the participants, day 3—departure to the territory of the Military Training and Research Center of the Ground Forces "Combined Arms Academy", which includes laying flowers on the alley of heroes, visiting the Museum of Military Glory and the exhibition of weapons: small arms, communication devices, military equipment. Then—a trip to the range with the performance of training shooting with Makarov pistol, machine gun Kalashnikov, machine gun, and sniper rifle. Shooting ends with a lunch from the field kitchen.

The winners and participants of the conference are awarded according to the results of the work of the sections.

4 Partnership

In connection with the targeted training the range of specialists, professors and teaching staff with the involvement of highly qualified scientific staff of SB RAS, such as the leading scientists of the Lavrentev Institute of Hydrodynamics, Gennady Anatolyevich Shvetsov, Vladimir Yuryevich Ulyanitsky, and the leading scientists of the Institute of Hydrodynamics of the Russian Academy of Sciences.The specialists from the applied research institute JSC "IPF" Polynovsky Anatoly Anisimovich, Chemeris Alexey Vasilievich, Braguntsov Egor Yakovlevich and others are also involved.

The department has prepared textbooks and manuals for the basic disciplines it reads [3–8].

In connection with the organization of cooperation and joint interaction on training highly qualified specialists under the program "New personnel for the high-tech industries", it is planned to organize a scientific and educational laboratory on the basis of the GDD department, which would be engaged not only in training personnel, but also in carrying out research work, both in fundamental and in applied tasks.

References

[1] I. Balagansky, L. Merzhievsky, A. Guskov et al., *Experimental and theoretical studies of high-speed processes* (NSTU Publising House, Novosibirsk, 2016)

[2] A. Guskov, K. Milevsky, *XVI All-Russian Scientific and Technical Conference "Science. Industry. Defense"*, Military Scientific Journal, **2**, 139 (2015)

[3] I. Balagansky, L. Merzhievsky, *Effects of weapons and ammunition* (NSTU Publising House, Novosibirsk, 2012)

[4] I. Balagansky, *Fundamentals of Ballistics and Aerodynamics* (NSTU Publising House, Novosibirsk, 2017)

[5] V. Andreev, A. Guskov, K. Milevsky, *Explosive substances* (NSTU Publising House, Novosibirsk, 2018)

[6] V. Andreev, A. Guskov, K. Milevsky, E. Slesareva, *Theory of combustion and explosion: high energy materials* (Urait, Moscow, 2021)

[7] M. Voronin, *Explosion and Impact Physics* (NSTU Publising House, Novosibirsk, 2019)

[8] A. Guskov, K. Milevsky, *Reliability of technical systems and technogenic risk* (NSTU Publising House, Novosibirsk, 2016)

Implementation of artificial intelligence development strategy in the Russian Federation in the educational programs of aerospace engineering training of Bauman Moscow State Technical University

Vladimir Kalugin[1], *Alexander* Lutsenko[1], *Irina* Romanova[1,*], and *Yongjian* Xu[2]

[1]Bauman Moscow State Technical University, 5/1, 2nd Baumanskaya st., Moscow, 105005, Russia
[2]Shanghai Jiao Tong University, 800 Dongchuan RD. Minhang District, Shanghai, China

Abstract. As a part of the National Strategy on Artificial Intelligence development in the Russian Federation, the specific features of training in USSN 24.00.00 Aviation and Rocket-Space Engineering at "Mechanical Engineering" (SM) Department of Bauman Moscow State Technical University were considered. The analysis of self-established educational standards and the classification of artificial intelligence course modules, implemented at the SM Department, is carried out. Individual courses that already use methods of artificial intelligence or are promising for the introduction of such methods are presented.

1 Introduction

Artificial Intelligence (AI) is a set of technological solutions that allows simulating human cognitive functions (including self-learning and searching for solutions without a predetermined algorithm) and obtaining results comparable at least to the results of human intellectual activity when performing specific tasks.

Although the first information systems for assisting people in making decisions date back to the mid-1950s, recent years have seen rapid growth in methods, algorithms, and practical applications of AI in various fields of activity.

Presidential Decree No. 490 of October 10, 2019 "On the Development of Artificial Intelligence in the Russian Federation" and the adopted accordingly National Strategy for the Development of Artificial Intelligence to the year 2030 (`https://www.garant.ru/products/ipo/prime/doc/72738946/`) determines both general tasks of the accelerated development of artificial intelligence in the Russian Federation and the major directions for enhancement of training system in this field. That includes conducting scientific research in the field of artificial intelligence and increasing the availability of information and computing resources for users.

At the Artificial Intelligence Journey (AI Journey 2020) conference on "Artificial Intelligence—Main Technology for the 21st Century" (`http://www.kremlin.ru/events/president/news/64545`), the goal for education was specified in a Presidential address: (quote) "I believe that training courses, modules on artificial intelligence and big data analysis should be included in educational programs, and for all areas of professional activity".

*e-mail: irina.romanova@bmstu.ru

Following the conference on Artificial Intelligence, the President approved the list of instructions, particularly according to Pr-2242, paragraph 1 e) —2: to complement the educational programs of higher education in all fields of specialization and areas of training with sections on the study of artificial intelligence for training in the application of such technologies in various fields of activity.

According to clause 45 of the National Strategy, the main directions of increasing the amount of qualified personnel for the Russian market of artificial intelligence technologies and the level of public awareness about the possible areas of use of such technologies are, inter alia:

Paragraph a)—development and implementation of educational modules as part of educational programs at all levels of education, advanced training, and professional retraining programs to provide citizens with knowledge, competencies, and skills in mathematics, programming, data analysis, machine learning, contributing to the development of artificial intelligence.

Herewith, the convergent knowledge provided through integration of mathematical, natural science, and social and humanities education is a priority for the development of promising methods of artificial intelligence.

Paragraph c)—improving the quality of mathematical and natural science education of students (within both primary and complementary educational programs), its integration with social and humanities education, creating conditions for attracting students to advanced training in these areas.

This article considers both the current state of training specialists for USSN (consolidated group of training areas and majors) 24.00.00 Aviation and Rocket-Space Engineering at the "Mechanical Engineering" (SM) Department of Bauman Moscow State Technical University, and the vectors of implementation of artificial intelligence development strategy in the curriculum for this program.

The promising methods of artificial intelligence focused on the creation of fundamentally new scientific and technical products, have been analyzed in advance. They include:

- autonomous task solving;
- automatic design of physical objects;
- automatic machine learning;
- information processing based on new types of computing systems;
- interpretive data processing;
- big data processing;
- intelligent data analysis;
- intelligent decision support.

The experience of training specialists in the field of aerospace technology is discussed in articles [1, 3]. The works [4, 10] are devoted to the problems and methods of AI, and the peculiarities of intelligent systems, as well as their application in the aerospace sphere.

The problem of including one or another AI technology in the educational process depends on the training program. The basis in the development of core professional educational programs (CPEP) for this major at Bauman Moscow State Technical University is the independently established educational standards (IEES).

2 The analysis of IEES implemented at "Mechanical Engineering" Department for the USSN 24.00.00 specialists where the methods, technologies, and examples of the application of artificial intelligence should be studied

At present, the specialists are trained in 7 sub-departments: SM-1 "Spacecraft and launch vehicles", SM-2 "Aerospace Systems", SM-8 "Launch Vehicle Systems", SM-12 "Technology of Rocket and Space Engineering" (24.05.01 program, Design, Production and Exploitation of Aerospace Systems); SM-3 "Orbital Mechanics and Flight Control" (24.05.04 program, Navigation and Ballistic Support of Space Applications); SM-5 "Autonomous Information and Control Systems"; SM-13 "Aerospace Composite Structures" (undergraduate and graduate training programs 24.03.01, 24.04.01, Rocket Systems and Cosmonautics).

Primarily, it is necessary to analyze the IEEE in terms of the areas of professional activity that they contain:

- 25 Rocket and Aerospace Industry (fields: engineering and technology related to design, testing, operation and effectiveness assessment of aerospace and rocket hardware, barrel and jet artillery, study of ballistic, aerodynamic and kinematic properties and features of professional objects, navigation and ballistic support of their application, dynamics and control of motion (flight), space flight control, ballistic design);
- 32 Aviation (fields: aerodynamic design of promising models of aviation and aerospace technology, ground and flight aerodynamic testing of models, mock-ups and full scale structures of manned and unmanned aircraft).

All promising methods of artificial intelligence should be included in the department's training programs.

An analysis of competencies embedded in the IEES implemented in the SM department was also carried out. In particular, the study of AI methods will contribute to the achievement of the following competencies:

- Fundamental Universal Competences (UCF): UCF-1 "Is capable of search, critical analysis and synthesis of information, its semantic optimization and visual presentation, apply a systematic approach to solve the tasks; use the basics of philosophical knowledge and analyze the patterns of the historical development of society to form worldview and civic position";
- Fundamental General Professional Competencies (GPCF): GPCF-1 "Is capable of applying natural science and general engineering knowledge, methods of mathematical analysis and modeling, theoretical and experimental research skills to solve various professional tasks"; GPCF-2 "Is capable of understanding the principles of modern information technologies and use them for analyzing and solving professional tasks"; GPCF-6 "Can develop physical and mathematical models of space and rocket-to-satellite objects"; GPCF-9 "Is capable to use scientific and creative skills to search and find non-trivial solutions of scientific and technical problems and create innovations in the rocket and space equipment design field";
- Fundamental Professional Competencies (PCF): PCF-1 "Is capable of forming the logic of spacecraft control system functioning in case of on-board equipment failures"; PCF-4 "Is capable of conducting research work, coordinating and managing the processes of developing the design and technological documentation at all stages of the aerospace product life cycle"; PCF-6 "Is capable of conducting scientific analysis and forecasting the requirements development for the GNSS means and tools and their functional additions".

3 Analysis of the implemented curricula and course programs of the SM Department for specialists in USSN 24.00.00 where methods, technologies, and examples of the application of AI are covered

The course programs of sub-departments at the Mechanical Engineering Department were analyzed. All students of the Mechanical Engineering Department in USSN 24.00.00 study in some form individual methods, modules, or full courses of Artificial Intelligence.

Table 1. Disciplines topics of the SM Departments at USSN 24.00.00 in which AI methods are to be studied and/or implemented

Sub-Department	Courses with AI implementation
SM-2	Automated layout of manned and unmanned spacecraft (SC) and systems. Fundamentals of parallel computing methods.
SM-3	Methods of control in spacecraft flight control. Methods of information analysis for spacecraft flight control. Analysis of processing and displaying information from the spacecraft. Objective and types of telemetry information processing. Method of two-level control. Method of intelligent analysis. Types of spacecraft states. Variants of a "fuzzy" situation development. Classification of an emergency situation (ES). Safety protocol in case of ES. The analysis of the modern world development of processing and displaying information from spacecraft. Features of information transfer. Main restrictions on control. Features of decision making when controlling the spacecraft. Formation of executive-level flight plans (priorities, compatibility, quality criteria). Basic algorithms of executive-level planning. Optimization of scientific experiments planning. Analysis of applicability of mathematical methods to solve the flight planning problem. Consideration of applicability of artificial intelligence methods in planning tasks. Random events, values, and processes in ballistics and airborne boost-glide systems. Incompatible and independent events. Formulas for total and inverse probability. Probability functions and densities, initial and central moments, cumulants, correlations, characteristic functions. Gaussian random variables, vectors, and processes. Markov processes. Tracking systems for randomly maneuvering targets. Stationary mode of tracking. Spectral density. Calculation of the tracking error variance. Analysis of the influence of transmissibility parameters on the accuracy of target coordinate estimates. Optimal filters. Ground-based automated spacecraft control system.
SM-5	Adaptive control in aerospace systems devices. Adaptation as a method of counteracting uncertainty in control and information processing systems. Methods of realization of the stochastic adaptive control. Identification-type adaptive systems. Discrete adaptive systems with an implicit reference model. Discrete adaptive systems with a configurable model of the control object. Statistical methods of identification. Regression analysis. Identification of a nonlinear system. Hammerstein's model. Measurement errors. Measurement function errors. Model errors. Accuracy assessment.

Table 1. (Continue)

Sub-Department	Courses with AI implementation
SM-5	Feasibility and repeatability of measurements. Optimal composition of measurements. The order of drawing up the program and methods of conducting tests of experimental and mock-up launchers during research and development work. Synthesis of optimum control systems based on dynamic programming method. The reduction of the optimal control problem to the mathematical programming problem by the method of projective matrix operators.
SM-8	Features and models of equipment operation in technical and launch systems based on random processes. Models equipment operation in technical and launch systems based on the mass service systems. Fundamentals of systematics. System analysis of ground equipment of aerospace systems. Fundamentals of the reliability theory. Calculation of basic fault tolerance, durability, and reliability indicators. Calculation of reliability indicators for various distribution laws of failure probability. Reliability of complex systems. Finite deterministic automata. Mile and Moore automata, combinatorial automata. Formal description of automata, automata tables, and graphs. Hardware and software implementation of automata.
SM-12	Acquiring the skills of practical and independent research of the effective operation of industrial processes based on the use of modern methods of quality control and analysis. "Statistica" program. Extracting knowledge from experimental data to obtain useful information and make decisions during the creation of numerical models of technological processes. Application of decision-making theory and expert judgment to select (and justify) a research direction. Methods for Finding Optimal Decisions in Selecting Test Options for Aerospace Equipment Products Based on Data Analysis (Properties and Operating Conditions of Rocket and Space Products). Optimization of technological parameters using analytical and numerical simulation of material processing processes based on the analysis of initial data. Decision-making methods for optimizing the variants of design technology. Development of expert and information retrieval systems to support the standard decision-making on the design technologies. Development of tests for students in the MS Forms environment with scoring in an automated mode based on the completed survey forms. Implementation of AI in the educational process).

Table 1. (Continue)

Sub-Department	Courses with AI implementation
SM-13	Methods for planning experimental studies and analysis of the reliability of the data obtained. Processing of experimental data to obtain composite materials properties. Using genetic algorithms for solving inverse problems. Analysis of random and methodical errors of temperature measurements during the bench tests of materials and structural elements. Application of machine learning algorithms for constructing multidimensional dependencies between operating parameters and calculated parameters of the thermal stress state of aircraft elements. Implementation of optimal solution search algorithms in Python environment. Automation of the design process in the Ansys environment using the Fortran/APDL programming language. The application of multilayer perceptron from the Python Sklearn library for constructing dependencies between input and output parameters of technical objects. Fundamentals of implementation of parallel computations in gas dynamics, heat transfer, and strength problems. The structure of modern computer-aided design systems. Statistical processing of experimental data. Testing statistical hypotheses, the choice of data distribution law. Application of planning algorithms for multifactor experiments (Python DOE library). Application of the high-level Pandas Python library for big data analysis. Series and Dataframe data structures.

The detailed analysis of Artificial Intelligence methods leads to the following classification of artificial intelligence course modules, already implemented in the SM department:

- "Artificial Intelligence Methods (General Course)" (7 courses);
- "Fundamentals of Data Analysis" (12 courses);
- "Data Analysis and Machine Learning" (10 courses);
- "Neural Networks" (5 courses);
- "Big Data Processing Systems" (3 courses);
- "Knowledge Engineering" (6 courses).

In each sub-department, among the studied programs, we found courses that can serve as a basis for the organization of intra- and interuniversity interaction for the preparation of new and modification of existing artificial intelligence curriculum.

Table 1 shows the most interesting topics of individual disciplines that already use methods of artificial intelligence or are promising for the introduction of such methods.

4 Conclusions

1. Sub-departments are divided into two groups:
   - ready-made courses in AI methods;
   - aspects of the application of AI methods in courses.

2. All departments are united by the availability of applied use of AI methods in specific majors and fields of training.
3. The direction of the curriculum improvement according to the National Strategy:
   - scientific and methodological elaboration of applied use of AI methods in the specific training field;
   - development of individual models, adjusted to the specifics of aerospace and defense engineering training;
   - experience exchange on the creation of educational and methodical sets, which consider separate aspects, general problems of Artificial Intelligence for non-profiled departments;
   - updating special courses on methods of artificial intelligence for majors that directly use these methods in their project activities;
   - retraining of faculty members in methods of teaching individual modules and courses in general in the field of artificial intelligence;
   - promoting scientific research of applied aspects of a particular focus;
   - integration of professional training and the economic cycle of disciplines with compulsory innovation components;
   - development of training, educational and methodological manuals on methods of artificial intelligence for aerospace and defense technology.

References

[1] V.T. Kalugin, *Faculty of Special Engineering MSTU named after N.E. Bauman—80 years*, Defence Equipment, **8–9**, 3 (2018)

[2] G. Shcheglov, V. Mayorova, V. Zelentsov, *Integration of Small Satellites Design Process into The Specialist's Degree Educational Program*, in Proceedings of the 69th International Astronautical Congress (IAC 2018), 1–5 October 2018, Bremen, Germany (2018)

[3] V.I. Mayorova, D.A. Grishko, V.V. Leonov, *Vivid mathematics as a general vector of multidisciplinary STEM education for future aerospace engineers*, Acta Astronautica, **178**, 72 (2021), DOI: 10.1016/j.actaastro.2020.09.003

[4] N.V. Lebedeva, S.V. Solov'yev, *Use of intelligent systems in operational control of spacecraft flight*, Aerospace MAI Journal, **25(2)**, 152 (2018)

[5] I.K. Romanova, *Study of the influence of deformations and asymmetry on aircraft movement parameters*, AIP Conf. Proc., **2171**, 130001 (2019)

[6] I.K. Romanova, *Application of data mining technologies in the tasks of detecting damage to unmanned aerial vehicles*, in Proceedings of the VII All-Russian Scientific and Practical Conference "Technologies for Development And Debugging of Complex Technical Systems", 1–2 April 2020, Moscow, Russia, 393 (2020)

[7] S.V. Solov'yev, *Justification of choice of clustering method and criterion for mining during spacecraft flight control*, Spacecrafts and Technologies, **4(33)**, 151 (2020)

[8] S.V. Solov'yev, *Principles for constructing an intelligent ground-based space flight control system*, Modern Science: actual problems of theory and practice. Natural and technical sciences, **3**, 144 (2021)

[9] I.K. Romanova, *Modern approaches to detecting changes in shape and other damage to flight vehicle during movement*, AIP Conf. Proc., **2318**, 110001 (2021), DOI: 10.1063/5.0036045

[10] N.A. Serdyukova, V.I. Serdyukov, *Smart System Sustainability and Smart System Management*, Intelligent Systems Reference Library book series, **191** (2020), DOI: 10.1007/978-3-030-54470-6_9

Development of teaching programs of artificial intelligence methods in aerospace education

Vladimir Kalugin[1], *Alexander* Lutsenko[1], *Irina* Romanova[1,*], and *Ding* Ye[2]

[1]Bauman Moscow State Technical University, 5/1, 2nd Baumanskaya st., Moscow, 105005, Russia
[2]Harbin Institute of Technology, 92, West Da-Zhi st., Harbin, Heilongjiang, 150001, China

Abstract. As a part of the National Strategy on Artificial Intelligence development in the Russian Federation, the basic concepts and definitions related to artificial intelligence (AI) and the necessity for the preparation of training programs were analyzed. The analysis of Federal state educational standards (FSES) and professional standards for enlarged groups of specialties and directions (EGSD) 24.00.00 AEROSPACE ENGINEERING was conducted and the perspective programs on the foundations of artificial intelligence and intelligent data analysis were proposed.

1 Introduction

The relevance of the considered task is determined by the Presidential Decree No. 490 of October 10, 2019 "On the Development of Artificial Intelligence in the Russian Federation".

The educational technologies should also rely on the Federal Law of 29.12.2012 No. 273-FZ "On Education in the Russian Federation", the Federal standards of higher education concerning the professional standards of the consolidated group of training areas and majors EGSD (enlarged groups of specialties and directions) 24.00.00 Aviation and Rocket-Space Engineering.

The application of artificial intelligence (AI) methods in education include two main trends:

- use of AI methods in the organization of the educational process;
- teaching of AI methods.

The first trends is universal. Technologies such as adaptive learning, personalized learning, automatic assessment, interval learning, evaluation of the faculty by students, smart campuses, control of the examination process are currently under development. Appropriate information platforms are also being developed for organizing the educational process using new technologies.

An enlarged classification of the AI technology application includes the following aspects:

- computer vision technologies in education (for monitoring students during online proctoring);
- technologies of natural language processing, speech recognition, and synthesis;
- technologies of intellectual support for decision-making.

The second trends of implementing AI methods in the curriculum relies on both the latest achievements in AI and the specifics of educational activities.

*e-mail: irina.romanova@bmstu.ru

In articles [1, 2] general approaches to the formation of educational programs are considered. The problems of digital transformation at a new stage are considered in [3, 4]. In [5–10], descriptions of methods of using intelligent technologies are given. Foreign experience is analyzed in the article [11].

2 Key concepts and definitions in the AI field used in the curriculum development

The key term, "artificial intelligence", is defined as "a set of technological solutions that allow the simulation of human cognitive functions, obtaining results comparable, at least, to the results of human intellectual activity".

An expanded definition of AI is as follows: AI is the science and technology of creating intelligent machines, especially intelligent computer programs. AI is related to the similar task of using computers to understand human intelligence but is not necessarily limited to biologically plausible methods. The goal of AI is to create technical systems capable of solving non-computational tasks and performing actions that require the processing of meaningful information and are considered the prerogative of the human brain.

The main attributes and distinctions of intelligent systems (IS):

1. Developed communicative abilities—processing arbitrary requests in a dialogue in a language closest to natural language (the system of natural language interface—SNLII).
2. Focus on solving weakly structured, poorly formalized problems (implementation of soft models).
3. Ability to work with undefined and dynamic data.
4. Ability to develop a framework and extract knowledge from the accumulated experience of specific situations.
5. The ability to obtain and use information that is not explicitly stored, but is derived from the database.
6. The system has not only a model of the subject area but also a model of itself, which allows it to define the scope of its competence.
7. Ability to draw conclusions by analogy.
8. Ability to explain its actions, user failures, to warn the user about some situations that lead to the violation of data integrity.

In contrast to conventional analytical and statistical models, IS allows obtaining a solution to poorly structured problems that are difficult to formalize.

The main types of intelligent systems are:

1. Systems with intelligent feedback and intelligent interfaces.
2. Computer-aided pattern recognition systems.
3. Computer-aided decision support systems.
4. Expert systems.
5. Neural networks.
6. Cognitive modeling.
7. Genetic algorithms and evolutionary modeling.
8. Knowledge extraction from experience (empirical facts) and Data Mining.

According to the above list, data mining is included in the list of intelligent systems. An alternative concept is the interpretation of Data Mining, which includes AI methods and is defined as a decision support process based on the search for hidden patterns in data.

The purpose of pattern finding is to present data as a representation of the searched processes. Building prediction models is also the goal of pattern search. Data Mining tools, unlike statistical tools, do not require a strictly defined amount of retrospective data. At the same time, it is necessary to control the statistical significance of the detected knowledge.

3 Analysis of FSES and professional standards for EGSD 24.00.00, and preparation of prospective courses on AI methods

Analysis of the FSES (Federal state educational standards) and professional standards combined with the AI methods and data mining was conducted to determine the focus of the individual modules and general courses development.

Seven FSES of 3++ generation were studied, and the majors for which it is feasible to include both individual modules and comprehensive training on AI methods were determined. Integrated training is recommended for the training programs 24.05.03 (Aircraft testing); 24.05.04 (Navigation and flight support of space vehicles); 24.05.05 (Integrated aircraft systems); 24.05.06 (Aircraft control systems).

In addition, 60 professional standards in the field of Rocket and Aerospace Industry were analyzed. AI General Course was recommended to be included in the training system for 22 of considered professional standards.

Analysis of 16 professional standards in the Aviation field showed the advisability of studying the general course of AI methods for at least four standards.

4 Development of the basic program on the Artificial Intelligence methods, recommended in the EGSD 24.00.00 educational programs

The main document to be developed is the educational program: a set of basic properties of education (volume, content, expected outcomes) and organizational and methodological conditions. This set is presented as a curriculum, a calendar of the curriculum, working programs of subjects, courses, disciplines (modules), other components, assessment and methodological materials, and in cases prescribed by Federal Law No. 273-FZ of 29.12.2012 "On Education in the Russian Federation", as a working program.

The following program is offered to form a general basic course on AI methods. The course consists of modules that may be considered both as a whole and as separate components.

The specifics of the course are fulfilled through the consideration of end-to-end examples in lectures and practical classes.

The objectives of the discipline include the formation of a system baseline representation and primary knowledge, skills, and competencies in the AI basics and methods, along with insights into the implementation of artificial intelligence methods in technical applications. This course prepares teachers and students to apply the concepts of intelligent systems in scientific and practical areas.

The main tasks of the discipline are to acquire skills in the research and development of artificial intelligence systems in engineering practice.

The specific problems of implementing artificial intelligence technologies depend on the focus of a particular educational program. Therefore, a significant element here is an educational module—a set of educational elements sufficient to organize and pass the training on a particular topic (class) of the discipline (course).

Table 1. Modules and contents of the general course on AI methods

Module	Topic
General issues of Artificial Intelligence	Lecture 1. General issues of Artificial Intelligence. Basic concepts and definitions of Artificial Intelligence. Place and role of the Artificial Intelligence Systems in Control Theory Applications
	Lecture 2. Models of cognitive psychology. Models of information perception in intelligent systems. Logic in Artificial Intelligence. Heuristic methods. Decision-making in an intelligent environment. Reasoning models. Dispute models. Prolog-logic programming language
Knowledge Engineering. Expert systems	Lecture 3. Representational languages and knowledge manipulation languages. Knowledge engineering. Intellectual knowledge bases. Knowledge representation models. Object-oriented databases. Evolutionary modeling
	Lecture 4. Expert systems. The concept of expert systems in control tasks for objects and processes in various activities
Mathematical basics of Artificial Intelligence methods	Lectures 5–7. Mathematical basics of Artificial Intelligence Methods. DATA Science, Statistics and Machine Learning. Clustering. Classification. Regression. Linear methods of classification and regression, nonlinear methods of classification and regression. Basics of Genetic Algorithms. Methods for model reconstruction from experimental data
Software implementation of Intelligent Systems	Lectures 8–9. Software implementation of Intelligent Systems. Basic algorithms of Machine Learning. AI languages and programs. Python basics. Lists, tuples, dictionaries, functions, lambda-functions, exceptions, plugins, classes
Neural Networks and Fuzzy Logic	Lectures 10–11. Neural networks (NN). Neural networks in control tasks. The basic architecture of artificial neural networks, the principle of training. Backpropagation method. The general scheme of training. Implementation of a simple neural network in Python. The architecture of networks: the choice of activation function, auxiliary layers, weights initialization. Parameters update, optimization methods. Methods for training neural networks and intelligent agents. Convolutional NN. Problems solved by convolutional NN. Introduction to modern architectures of NN: VGG, GoogleNet, LeNet. Reinforcement learning. Functions for describing agent's behavior. Deep Q Learning algorithm. Open AI Gym library. Basic features of the Neural Toolbox package of Matlab

Table 1. (Continue)

Module	Topic
Neural Networks and Fuzzy Logic	Lecture 12. Fuzzy logic methods. The basic concept of fuzzy sets. Methods for defining membership functions, their analytical and graphical representation. Methods of defuzzification. Fuzzy inference systems
	Lecture 13. Pattern recognition. Image analysis. Application of traditional technologies and Artificial Intelligence technologies in solving recognition, classification, and ranking tasks of analysis objects
Intelligent Control Systems	Lecture 14. Intelligent Control Systems. Structural diagrams using neural controllers. Structural schemes of a regulator with fuzzy logic. Complexation of neuro-fuzzy robust control methods. Problem statement for the complexation of readings of heterogeneous sensory systems
	Lecture 15. Intelligent robotic systems
Application of AI methods in aerospace engineering	Lectures 16–17. Applied problems of implementation of AI methods for a specific training program: aerospace engineering

Table 2. DATA Mining course syllabus

Module	Lecture	Seminar
Developing models and generating data for research	Data. Data operations. Dirty data processing methods	Two types of models: Big Data and limited data. Simulink models and system of ordinary differential equations models for end-to-end examples (on the example of the aircraft). Technologies of parallel computing in MATLAB
Dirty data processing methods	Tasks, methods, and stages of DATA Analysis—DATA Mining (DA-DM)	Work with missing data. Handling inconsistent data. Data filtering. Data smoothing with convolution. Removing the Data Trend from the Experimental Data Processing for Cross-Cutting Examples (using aircraft as an example)
Statistical methods in DATA Mining	Statistical methods in DA-DM. Distribution Forms	Key Indicators. Descriptive Statistics. Calculating and plotting descriptive statistics. Example 1—Calculation of Maximum, Mean, and Standard Deviation. Example 2—subtracting the mean using MATLAB data statistics for end-to-end examples. Distribution Shapes. One-Dimensional Histograms. Two-dimensional Histograms Multivariate Histograms for describing the quality indicators of the cross-cutting examples

Table 2. (Continue)

Module	Lecture	Seminar
Regression analysis	Data transformation and Dimensionality reduction (linear and nonlinear methods)	Linear Correlation. Linear Regression. Simple Linear Regression. Residuals and quality of the approximation. Example: Calculating R2 with Polynomial Approximations. Calculation of Adjusted R2 for Polynomial Regressions. Basic Fitting UI. Preparing for basic fitting. Opening Basic Fitting UI. Interpolating and extrapolating values. Creating a code file to display the result. MATLAB functions for polynomial models
Data Conversion and Dimensionality reduction	Time Series Analysis	Data Conversion. Dimensionality reduction-Linear methods. Introduction. Principal Component Analysis PCA
Time Series Analysis	Clustering	Creating and modifying time series objects. Viewing Objects in a Time Series. Changing Time Series Objects and Interpolation Method. Event detection. Creation of objects in a collection of time series. Removal and interpolation of missing data. Display time vector values as a string of data. Creation of Plots of Time Series Collection Members. Time series generator
Clustering	Classification and Regression Smoothing	Hierarchical Methods. Optimization methods. K-means method for evaluating the impact of system structure and parameters on the quality performance of the end-to-end example
Classification and regression	Graphical methods and cluster visualization. Multivariate visualization. Data tours. Interactive data exploration	Classification rule construction methods. Decision Tree methods. Methods for constructing mathematical functions. Regularization networks. Discretization and sparse grids for end-to-end examples. Prediction of time series
Multivariate visualization	Factor analysis	Multivariate visualization Factor analysis Glyph charts. Scatterplots charts. Scatterplot Matrices. Hexagonal scatterplots charts. Dynamic Graphics. Data identification. Linking. Coplots. Dot Diagrams. Andrews Curves

The educational module outcome is the development of certain knowledge, abilities, skills, or competencies by the student.

Description of the modules of the general course on AI methods is presented in table 1. According to the analysis of FSES 3++ for EGSD 24.00.00 and relevant professional standards in the fields of Rocket and Aerospace Industry and Aviation, we recommend including the "Intelligent Analysis of Data—DATA Mining" course for all training programs.

DATA Mining skills will become the basis in solving the most important problems for all engineers designing new equipment, specialists involved in the equipment operation, research engineers, and scientists. The received knowledge and skills will be irreplaceable for design decision-making support, prognosis, choosing the development directions, Big Data processing, and problem-solving in the limited information conditions.

An analysis of the problems in the aerospace industry resulted in the following course on data mining, presented in table 2.

5 Conclusions

1. The analysis of FSES 3++ on EGSD 24.00.00 "Aviation and Rocket-Space Engineering" and professional standards in the field of Rocket and Space Industry and Aircraft Engineering was performed in terms of the necessity to include the study of AI methods in professional training.

2. The courses and individual modules on the basics of AI and data mining methods were developed for implementation in the curriculum of the EGSD 24.00.00.

3. Recommendations on the implementation of both general courses and individual modules of AI methods into the training system of specialists have been prepared.

References

[1] O. Oreshkina, V. Aslamazova, *Specifics of developing and implementing remote classes in chemistry with hearing impaired students at a technical university of a general type*, IEEE Global Engineering Education Conference (EDUCON), **4**, 244 (2021), DOI: 10.1109/EDUCON46332.2021.9454033

[2] A.L. Galinovsky, M.I. Abashin, A.A. Abashina, A.S. Vyshegorodtseva, *Modem models of engineers' training in Russia*, AIP Conf. Proc., **2318**, 150005 (2021), DOI: 10.1063/5.0035807

[3] I.K. Romanova, *Methodological aspects of application of software in teaching engineering disciplines: tasks, problems and perspectives*, in International Forum "IT-Technologies for Engenering Education: New Trends and Implementation Experience": Conference Proceedings, 28–29 November 2019, Moscow, Russia, 339 (2020)

[4] N. Vatolkina, M. Dos Santos Cardoso, *Management of digital transformation of educational technology: Key elements*, in International Symposium on Project Approaches in Engineering Education; Active Learning in Engineering Education Workshop; International Conference on Active Learning in Engineering Education (PAEE/ALE'2021), 7–9 July 2021, Braga, Portugal (2021), DOI: 10.5281/zenodo.5095070

[5] Y.I. Dimitrienko, E.A. Gubareva, *Neural network model of mathematical knowledge and development of information and educational environment for mathematical training of engineers*, J. Phys.: Conf. Ser., **1141**, 012010 (2018), DOI: 10.1088/1742-6596/1141/1/012010

[6] O. Serebrennikov, A. Ponomarev, A. Proletarskiy, G. Guglya, *Usage of artificial intelligence in quality improvement of educational programs*, in eLearning and Software for Education Conference, 16th International Scientific Conference on eLearning and Software for Education (eLSE 2020), 476 (2020), DOI: 10.12753/2066-026X-20-148

[7] N.A. Serdyukova, V.I. Serdyukov, S.S. Neustroev, E.A. Vlasova, S.I. Shishkina, *Smart Algebraic Approach to Analysis of Learning Outcomes*, Smart Innovation, Systems and Technologies, **188** (2020), DOI: 10.1007/978-981-15-5584-8_41

[8] L.V. Juravleva, V.A. Shakhnov, A.I. Vlasov, *Adaptation of professional engineering training to the challenges of modern digital production*, Advances in Intelligent Systems and Computing, **1134** (2020), DOI: 10.1007/978-3-030-40274-7_59

[9] S.S. Sokolov, M.N. Saveleva, A.V. Mitrofanova, S.V. Kolesnichenko, N.S. Logunov, *Implementation of Training Programs Using Digital Distance Education Technologies for Seafarers*, in Proceedings of the 2020 IEEE Conference of Russian Young Researchers in Electrical and Electronic Engineering (EIConRus), 27–30 January 2020, St. Petersburg and Moscow, Russia (2020), DOI: 10.1109/EIConRus49466.2020.9039034

[10] N.G. Bondarenko, G.I. Lukyanov, T.M. Makhamatov, L.I. Egorova, T.I. Avdeeva, E.N. Belyanova, *Education and information technology: Blended learning and inquiry-based learning innovative solutions*, Journal of Advanced Research in Dynamical and Control Systems, **12** (2020), DOI: 10.5373/JARDCS/V12SP1/20201086

[11] A.Yu. Lutsenko, I.R. Shafikova, A.L. Galinovsky, *Training features of the engineering personnel for the high-tech sector of the economy in the UK*, HANDBOOK. An Engineering journal with appendix, **10**, 50 (2017), DOI: 10.14489/hb.2017.10.pp.050-054

 https://doi.org/10.1051/shsconf/202213701007

Analysis of the impact of the introduced restrictions caused by the new COVID-19 infection in the field of education on the results of the unified state exam in various constituent entities of the Russian Federation

Ilia Konstantinov[1,*], *Anna* Zhuleva[1], and *Denis* Shavelkin[1]

[1]Department 101 "Design and Certification Aviation Technology", Moscow Aviation Institute (National Research University), 4, Volokolamskoe shosse, Moscow, 125993, Russia

Abstract. Since the end of 2019, all countries of the world have faced a rapidly increasing level of disease caused by the new coronavirus infection COVID-19, which in turn has led to the introduction of a number of restrictions in various areas of life. Educational institutions in many countries were forced to transfer students to a remote mode of work, for which both sides were not ready. The difficulty of a quick transition to a distance learning mode was primarily associated with a number of technical limitations, since not every educational institution had the proper material and technical support, just like not every student had the opportunity to use electronic resources. The resulting situation temporarily made it impossible to carry out the educational process. As technical and other problems were solved, educational organizations sought to bridge the gap in the quality of distance and face-to-face training. This article provides statistical data on the results of the unified state exam in mathematics in various regions of Russia and analyzes the results of the exam.

Distance learning is a rather multifaceted aspect that cannot be unequivocally classified as a positive or negative phenomenon. Like any complex measure, it has both positive and negative aspects.

The clear advantages of distance learning include:

- The ability to conduct training regardless of the epidemiological situation in any subject of the Russian Federation or another country in the world [1]. Distance learning allows you to organize various forms of classes, including project activities, which contribute to the development of leadership qualities and skills within team communication with elements of the competition model [2].
- Distance learning is inextricably linked with the use of digital platforms, which allow high-quality storage, systematization and accumulation of educational material. To refer to this or that information or materials, a student only needs to have access to the Internet [3].
- Another equally significant advantage in the implementation of the distance educational process is the ability to organize monitoring of students' progress by the administration of educational institutions. Since electronic educational platforms often have a fairly wide functionality, it is also possible to check assignments in them, which allows you to track

*e-mail: ilyakonstantinov39@gmail.com

the teacher's statistics in real time throughout the course. The same category of advantages includes the fact that the teacher can track the student's activities outside the classroom, that is, watch when and for how long the student turned to various sections of the course [4].

- In terms of the implementation of international educational programs of higher education, it is required to attract experienced specialists from the industry, to create an infrastructure for the exchange of knowledge. The system should promote the development of journalistic activity of students and teachers. Remote mode leads to the development of this kind of activities. Thus, it becomes possible to attract world-class faculty to work, to create international conferences. It is noteworthy that participation in these events does not require distraction from daily routine affairs with a colossal expenditure of money and time resources.

However, along with the large number of positive aspects of introducing distance learning, there are a number of disadvantages. The obvious disadvantages include the following:

- High complexity in the transformation of educational and other activities caused by a significant load on the material and technical base of the educational institution. Moreover, the increased share of digital technologies in the activities of educational organizations demanded other new competencies from persons involved in the educational process [5]. Taking into account the education sector, as rather conservative, and due to the average age of the teaching and administrative staff, we can say that not all educational institutions have coped with the sharp digital transformation [6].
- Another, no less important disadvantage of the distance educational process is the complexity, and sometimes the impossibility of organizing distance events for people in different time zones. There are frequent cases in various international and network educational projects when classes for students were held with a difference of 5 hours or more. Obviously, this factor quite strongly affects both the degree of assimilation of the material and the degree of student involvement.
- Another difficulty of the distance form of work of an educational organization is the lack of regulatory and legal documentation for holding graduation qualification events for students in the framework of the State Final Attestation. The mechanism for identifying the student's personality and the legitimacy of this procedure is not fully understood. In this regard, most of such events in the Russian Federation were held in person, in compliance with all sanitary and epidemiological standards.
- Significantly increased complexity of the student's control during classes. Distance learning is not so critical for individuals with a fairly high degree of self-organization. While for students with low academic responsibility, the introduction of distance learning can be a relaxing factor [7, 8].

Thus, distance learning forms have both a number of positive and negative sides [1]. For a more objective assessment of the distance educational process, a number of studies are required with the distribution of weights by various factors. Another method of assessment is the statistical method, on the basis of which it is possible to form a method for forecasting and correcting the further development of educational activities.

Distance education has undoubtedly influenced the quality of mastering the educational program of students regardless of their level of education (schoolchildren, students) [9]. For the purpose of the analysis, the results of the Unified State Exam in Mathematics in various regions of the Russian Federation were taken:

- the city of federal significance—Saint-Petersburg (population 5.38 million) [10];
- the region of Kabardino-Balkaria (population 0.87 million) [10].

The above two constituent entities of the Russian Federation have different population sizes, and the average per capita expenditures of the region for the organization and provision of various institutions of the educational industry also differ. An important difference is the average income of a resident and the share of GDP that is formed by this subject [10].

The average per capita income in Saint-Petersburg is 2.4 times higher than the same in Kabardino-Balkaria, and the gross regional product per capita in Saint-Petersburg is 4.62 times higher than the same indicator in Kabardino-Balkaria. At the same time, the regional costs for the use and implementation of digital technologies differ 400 times in favor of Saint-Petersburg [10].

Kabardino-Balkaria

As can be seen from the table (table 1), in the subject under consideration, over 90% of students pass mathematics. In 2020, there is a significant decline in exam passing, which is primarily caused by the demographic crisis of the early 2000s in the region.

Table 1. The number of participants in the mathematics examination in Kabardino-Balkaria (for the last 3 years) [11]

Subject	2019		2020		2021	
	number of people	% from the total number of people	number of people	% from the total number of people	number of people	% from the total number of people
mathematics	8072	99.9	4363	93.3	7792	97.2

A difficult situation with coronavirus infection in the Russian Federation arose in the spring of 2020, it is this period of the year that is traditionally associated with the active phase of preparing students for the upcoming exams. Results distribution is shown in table (table 2).

Table 2. Dynamics of the results of the examination for the last 3 years in Kabardino-Balkaria in subject mathematics [11]

Criterias	Kabardino-balkaria		
	2019	2020	2021
Got low results	79	729	205
Average % of completion	58.5	42.6	50.2
Got high results	40	10	25
Got 100%	1	0	0

Undoubtedly, a significant increase in low scores for the exam (almost 10 times) is associated with insufficient digitalization of the region and the rather low speed of the subject's response to digital transformation. In view of the above, some of the students temporarily lost the opportunity to prepare for exams. The relatively low level of well-being of citizens living in this region aggravated the current situation. Due to the lack of modern telecommunication facilities for some students and the inability to use commercial educational services during the active phase of preparation for exams, the level of preparation for exams has significantly decreased.

Table 3. The main results of the examination in subjects in Kabardino-balkaria (mathematics) [11]

Result	0–10	11–20	21–30	31–40	41–50	51–60	61–70	71–80	81–90	91–100
Number of participants	83	87	473	1371	1889	1850	1224	615	190	10

A detailed distribution of student results is presented in the table (table 3).

The total number of those taking the exam in 2021. amounted to 7792 people. The results of table (table 3) can be shown in more detail in the diagram (figure 1).

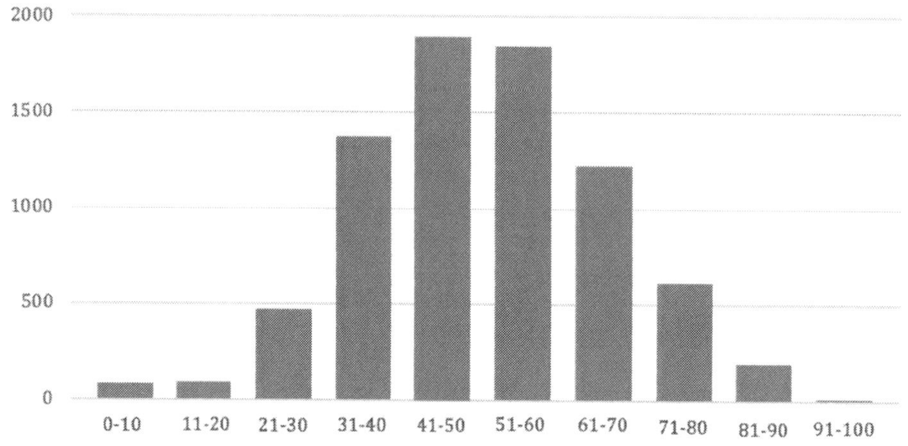

Figure 1. Distribution of the test scores of final state exam participants in the subject in 2021 in Kabardino-Balkaria

It can be seen from the above table (table 3) that the median value of the average exam result is concentrated in the range from 31 to 70 points, which is a relatively low result.

Saint-Petersburg

Table 4 shows that the percentage of exam participants over the past three years is concentrated in the 50% range. of the total number of exam participants. This turnout of graduates is primarily associated with the presence in the region of specialties in higher educational institutions that do not require the subject of "Mathematics" for admission.

Table 4. Number of participants of the Unified State Exam in the academic subject in Saint-Peterburg (for the last 3 years) [12]

2019		2020		2021	
number of people	% from the total number of people	number of people	% from the total number of people	number of people	% from the total number of people
16404	52.19%	16700	49.97%	18265	50.21%

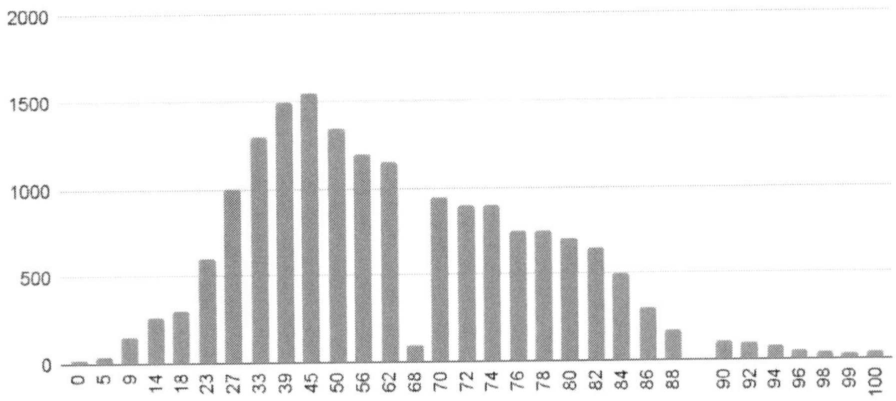

Figure 2. Distribution of the test scores of final state exam participants in the subject in 2021 in Saint-Petersburg

The diagram (figure 2) below shows that the median mean is concentrated in the range from 33 points to 56 [12].

The table below shows that the percentage of participants who did not score the minimum score in 2020 increased significantly compared to 2019, where no coronavirus restrictions were introduced.

The increase in the number of applicants who did not pass the minimum threshold is due to the same reasons as in another considered subject of the Russian Federation.

However, in view of the rather high level of digitalization in this region, the higher standard of living of the population, this increase is not so critical, in comparison with other considered subject.

In both regions under consideration, the results of 2021. They have improved for the better, this is due to the adaptation of the regional authorities to the changed conditions of the organization of the educational process. However, the results of 2021.

Still below the 2019 results (table 5). It is impossible not to mention the fact that in both regions colossal work was carried out to adapt educational institutions to distance learning.

Table 5. Dynamics of the final state exam results for the last 3 years in Saint Petersburg [12]

Criterias	Saint-Petersburg		
	2019	2020	2021
Did not get minimum results	4.70%	7.60%	7.71%
Average test result	58.89	56.16	56.24
Got from 81 to 99 points	10.80%	8.58%	10.54%
Got 100 points, people	86	23	25

Exam participant 2021 We spent a long time on distance learning in both the 10th and 11th grades, but thanks to the measures taken by the self-government bodies, we managed to stabilize this situation and neutralize the negative impact of distance learning.

Conclusion

From the data presented above, it can be seen that the new coronavirus infection significantly influenced the results of the Unified State Exam in Mathematics. A similar situation is observed not only in the subject "Mathematics", but also in all other subjects from the list of available for delivery. The Unified State Exam is an objective measure of comparing the quality of the educational process and the results of its development in different regions of the country, in view of the fact that these materials are developed by leading educational and methodological specialists at the Federal and regional levels. From the materials presented above, it can be seen that the mobility and flexibility of local government authorities, as well as the level of financial and personnel well-being of the region as a whole, have a significant impact on the result of passing the exam, as well as the digitalization of the population in the given subject.

References

[1] E.A. Volzhenina, V.A Kaigorodova, D.S. Shavelkin., *Methods of conducting classes in a distance learning environment*, Modern Pedagogical Education, **10**, 58 (2021)

[2] I.A. Konstantinov, V.A. Goncharova, *Competition Model in the Russian Higher Education System*, Education Management Review, **3**, 111 (2021)

[3] D.A. Prokopenko, V.A. Kaigorodova, *Aerospace education in Russia*, Staffing of the military-industrial complex, **11**, 164 (2021)

[4] S. Sengupta, *Possibilities and Challenges of Online Education in India During the COVID-19 Pandemic*, International Journal of Web-Based Learning and Teaching Technologies, **17(4)** (2022), DOI: 10.4018/IJWLTT.285567

[5] E.A. Gnatyshina, N.V. Uvarina, A.V. Savchenko, *Characteristics of adaptation processes at the university in the conditions of socio-cultural dynamics: Comparative analysis of adaptation processes of foreign and domestic university students*, Bulletin of the South Ural State University, **10**, 34 (2018)

[6] D.A. Prokopenko, O.S. Dolgov, B.B. Safoklov, D.B. Sukhanov, *Adaptation of a student of a higher educational institution to the educational process in the conditions of distance learning*, in Collection of the 9th International Scientific and Practical Internet Conference "Modern Challenges and Actual Problems of Science, Education and Production", **1**, 476 (2020)

[7] E.L. Bolotova, A.N. Rizhov, G.M. Kojaspirova, *Pedagogy of higher education* (Prospekt, Moscow, 2021)

[8] I. Gan, R. Sun, *Digital Barriers and Individual Coping Behaviors in Distance Education During COVID-19*, International Journal of Knowledge Management, **18(1)** (2022), DOI: 10.4018/IJKM.290023

[9] A. Parmar, A. Singh, *An Evaluation of Online Education Efficacy and the Measures to Improve It From the Perspectives of Management Students in the Wake of COVID-19*, International Journal of e-Collaboration, **18(1)** (2022), DOI: 10.4018/IJeC.290300

[10] S.M. Okladnikov et al., *Regions of Russia. Socio-economic indicators. 2020: Statistical Collection* (Rosstat, Moscow, 2020)

[11] The report on the main results of the USE in the subject—mathematics, prepared by the Department for the assessment of professional skills and qualifications of teachers is a structural subdivision of the State budgetary institution "Center for the assessment of the quality of education, professional skills and qualifications of teachers" of the Ministry of Education, Science and Youth Affairs of the Kabardino-Balkarian Republic (GBU "TSOKOPMK" of the Ministry of Education of the KBR, Nalchik, 2021)

[12] Results of the unified state exam in mathematics in 2021 in Saint-Petersburg. Analytical report of the subject commission (GBU DPO "SPbTSOKiIT", Saint-Petersburg, 2021)

Methodology for identifying the effective proportion of educational activities for the purpose of conducting distance learning without losing the quality of the educational process

Ilia Konstantinov[1,*], *Anna* Zhuleva[1], and *Denis* Shavelkin[1]

[1]Department 101 "Design and Certification Aviation Technology", Moscow Aviation Institute (National Research University), 4, Volokolamskoe shosse, Moscow, 125993, Russia

Abstract. The pandemic of the new coronavirus infection has significantly affected all areas of society in all countries of the world, the education sector is no exception. In view of the rapid spread of the growth of morbidity, many universities were forced to abandon the traditional educational model and switch to a remote mode of operation. However, such a transition raises a serious question about the quality of training of students. This paper analyzes the effectiveness of distance learning in one of the technical specialties of the Moscow Aviation Institute (National Research University). This analysis contains various statistical data. This study shows that although distance learning forms have a number of positive aspects, they also have a number of negative factors, in view of which a complete transition from the classical educational model to the distance form is impossible.

At the end of 2019, society faced a serious challenge that faced all spheres of life caused by COVID-19. In view of the intensive spread of the disease among various groups of the population, the main goals of the ministries of health in different countries of the world were to stop the increase in the incidence by limiting physical contact between people. So, many areas of business are faced with the need to carry out their activities online, education is no exception [1]. COVID-19 has brought significant changes to the way higher education institutions operate [2]. Also, when implementing distance learning forms, it is required to maintain the unchanged quality of the educational process, namely, to ensure the implementation of all the basic competencies that must be acquired by the student.

In the Russian system of higher education (Bachelor's, Specialist, Master's, Postgraduate, Doctorate) in various areas of training, a special set of competencies has been developed and monitored at the state level, which the student should acquire during the development of the educational program. These competencies, along with their development results, are spelled out in the Federal State Educational Standard [3], and the following groups of competencies are distinguished for the direction of training in the 24th group "Aviation":

- universal;
- general professional;
- professional;
- additional professional competencies.

*e-mail: ilyakonstantinov39@gmail.com

In turn, for each of the given groups of competencies, as well as for each competence, the results of their development are specifically presented, which can be divided into the following groups:

- to know;
- to be able;
- to have skills.

Schematically, these learning outcomes can be shown in figure 1.

The system of monitoring the activities of higher educational institutions adopted in the Russian Federation, as well as a number of standards developed for the implementation of educational activities, make it possible to make a fairly clear division of educational activities that pursue the achievement of various results of development. In the system of Russian higher education, the following forms of educational activity can be distinguished:

- lectures;
- practical classes;
- laboratory classes;
- student personnel scientific work;
- practice;
- state final certification.

Schematically, the above types of load can be represented on the diagram as follows (figure 2).

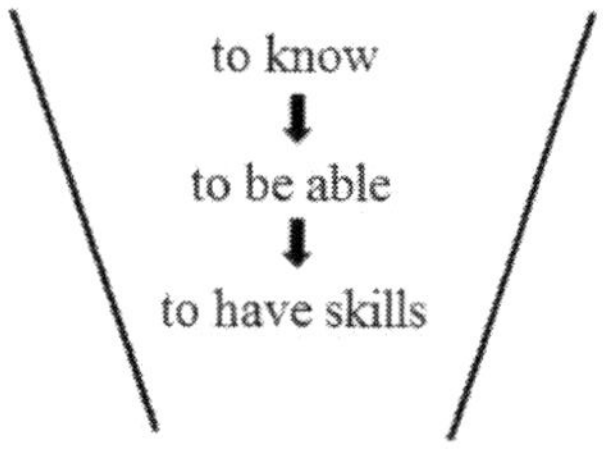

Figure 1. Learning outcomes

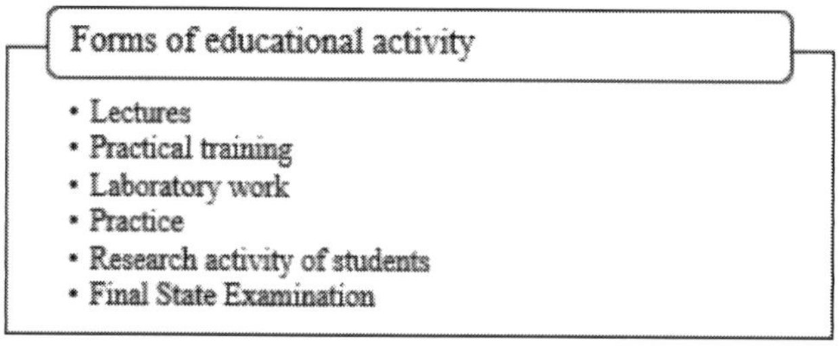

Figure 2. Forms of educational activity

Analysis of the effectiveness of the introduction of distance learning should not be attributed to the entire educational program as a whole; it is more correct to consider distance learning in the ability to achieve the required results of mastering each of the competencies presented.

This paper will consider the positive experience of introducing distance learning at the Moscow Aviation Institute.

Further consideration of the educational process will be based on the classification of competencies and the results of development, which were given earlier.

According to the legislation of the Russian Federation, the duration of the bachelor's educational program is 240 credits.

Structure of the Credit can be represented in formula (1)

$$\text{Credit} = T_{\text{classrom}} + T_{\text{student self}} = 36 \text{ hours.} \tag{1}$$

The structure of each credit unit within the framework of each specific course is a floating value, but with a sufficiently high accuracy it can be assumed that the number of credit points is divided equally between classroom and independent student work for 18 hours.

To assess the importance of taking certain measures, it is necessary to analyze in more detail the structure of the basic curriculum in the direction of "Aviation". So, according to this plan, 47.55% of the load is implemented in the format of lectures. The total number of classroom sessions is approximately 9000 hours. The remaining load can be divided as follows:

- 35% classroom load falls on practical classes;
- –7.45% of the classroom load falls on laboratory work.

Other types of load that are not taken into account in the classroom include:

- practices;
- final state certification.

Today, using modern digital platforms, it becomes possible to conduct lectures without losing quality. In order to improve the quality of the educational process, the lecture material can be recorded in advance, in which case the student will have the opportunity to view the material in good quality. The introduction of such activities can significantly reduce labor costs for the implementation of the educational process. The freed-up time of the teaching staff can be used to provide consulting support to students, as well as to modernize existing educational programs [4].

Modern educational platforms have a fairly wide functionality for conducting practical exercises of varying complexity and focus [5].

In the technical specialties of Russian higher education, the most popular types of practical training are:

- Project activities of students. Students are divided into groups of 3–7 people, assign roles in the team and solve a case. This form of the educational process has a beneficial effect on inter-student communication, which is especially urgently needed during physical restrictions on the interaction of students. Also, such type of educational activity allows to implement competitive model in the educational process [6].
- Another, no less popular format of practical training is the analysis of typical tasks, the purpose of which is to consolidate the theoretical material.

Both of these formats can be implemented online without loss of quality. However, despite all the advantages of distance learning, they have a number of undeniable disadvantages. It is impossible to carry out laboratory work with the help of only remote forms, since this format of conducting classes requires the availability and access to it of special equipment, which is located on the territory of the university.

Thus, based on the structure of a typical curriculum, approximately 83% of the classroom load can be carried out in a distance format.

It is necessary to analyze in more detail the structure of the remaining 17%, which are implemented in the form of laboratory work.

Laboratory work, depending on the nature of the subject and the competence acquired, can be one of the following types, shown in figure 3:

- Computational—graphic laboratory work. As part of the implementation of this type of laboratory work, the student, using modern software, learns to solve typical problems of programming and engineering. The difficulties of organizing this type of classes include the fact that, depending on the specific software manufacturer, it is not always possible to install software on a student's personal computer. However, most software vendors have student licenses. Another equally significant drawback is the fact that even with student software licenses, this imposes additional requirements on the technical capabilities of a student's personal computer.

- Technological laboratory work. This type of laboratory work is most common in disciplines that have the goal of studying technological processes and/or studying industrial equipment that is used in various enterprises.
- Laboratory work for the study of fundamental sciences. In the direction of training "Aviation", a fairly significant share of competencies is devoted to the study of general engineering subjects. These subjects include: physics, chemistry, resistance of materials, etc. These laboratory work should also take place in specialized classes, using the required equipment in compliance with all safety measures.

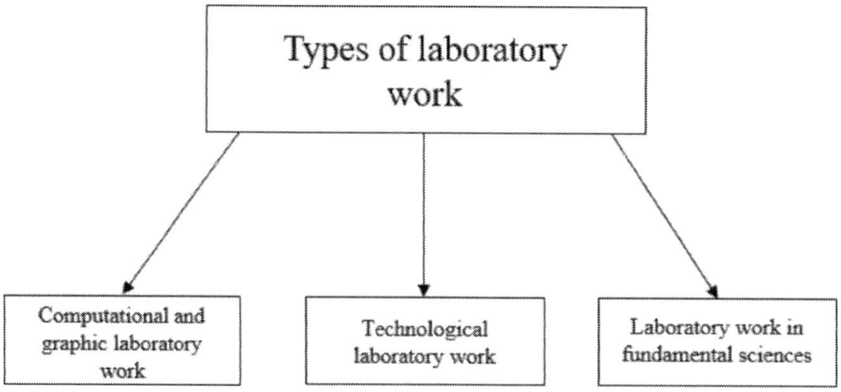

Figure 3. Types of laboratory work

As part of the implementation of the study under consideration, the percentage of laboratory work of the first type to all possible laboratory work is approximately 64%. Thus, 64% of 17% of implemented laboratory work can be carried out remotely

$$N_{\text{lab work dist}} = 0.64 \cdot 0.17 \cdot 100\% = 10\%. \quad (2)$$

Consequently,

$$N_{\text{dist}} = 83\% + 10\% = 93\%. \quad (3)$$

Thus, approximately 93% of educational events held at the Moscow Aviation Institute (National Research University) can be implemented remotely without losing the quality of the educational process. However, within the framework of this study, other factors that may affect the assimilation of certain competencies were not taken into account. One of the most significant factors is the ability of students to organize themselves. It turns out that 93% of educational activities do not entail a loss of quality for students who are conscientious about the educational process [7, 8].

Based on the structure of the basic curriculum and taking into account the total number of classroom sessions, it is not difficult to calculate the weighted average face-to-face load for students and faculty for laboratory work.

Thus, assuming that laboratory work is distributed evenly over the years of study, and also evenly distributed within each academic year, we get:

$$T_{\text{offline lab per week}} = \frac{9000 \cdot (1 - 0.93)}{4 \cdot 35} = 1.16 \text{ groups work/week.} \quad (4)$$

Where 35 is the average number of academic weeks per year, and 4 is the duration in years of the "Bachelor's" education program.

At the moment, the number of undergraduate students of the Moscow Aviation Institute is just over 10,000 students, and the average size of the study group is 12 people. Thus,

$$N_{\text{groups}} = \frac{10000}{12} = 833. \tag{5}$$

And the total load of the university for laboratory work is

$$T_{\text{offline lab}} = 833 \cdot 1.16 = 997 \text{ lab hours/week.} \tag{6}$$

With an average number of faculty members of the Moscow Aviation Institute $N_{\text{prof}} = 3000$ we will get the average load on the teacher for conducting face-to-face laboratory work per week

$$T_{\text{prof lab}} = \frac{997}{3000} = 0.33 \text{ prof hour/week.} \tag{7}$$

Laboratory work, depending on the specifics of the course and its duration, has a duration of 2 or 4 academic hours, respectively, respectively, each teacher can conduct laboratory work for a group once every 6 weeks and 12 weeks, respectively, which in turn fully meets the sanitary and epidemiological standards and exceeds the quarantine period of 14 days, which, to a fairly high degree, can reduce the growth of morbidity among students.

Conclusion

This paper provides a detailed analysis of one of the main curricula implemented at the Moscow Aviation Institute. It was shown that most of the educational load in the amount of 93% of the entire classroom load can be implemented remotely without losing the quality of the educational process and without distorting the requirements for the results of mastering each specific competence.

As noted in this study, the achievement of such high values of the possible number of educational activities implemented with the use of digital technologies is possible only with a sufficiently high level of self-organization of students and requires quite significant efforts from the administrative staff to adapt the educational process to rapidly changing realities.

One of the most significant conclusions of this work is the fact that in view of the relatively low percentage of the classroom load, which can be implemented exclusively in full-time work, it was shown that such laboratory work can be carried out in compliance with all measures to prevent an increase in morbidity among students.

References

[1] I. Gan, R. Sun, *Digital Barriers and Individual Coping Behaviors in Distance Education During COVID-19*, International Journal of Knowledge Management, **18(1)** (2022), DOI: 10.4018/IJKM.290023

[2] D.A. Prokopenko, V.A. Kaigorodova, *Aerospace Education in Russia*, Staffing of the military-industrial complex, **11**, 164 (2021)

[3] Order of the Ministry of Education and Science of the Russian Federation of February 5, 2018 No. 81 "On the approval of the federal state educational standard of higher education—bachelor's degree in the direction of training 03.24.04 Aviation" (as amended and supplemented), Edition as amended by No. 1456 of November 26, 2020

[4] D.A. Prokopenko, O.S. Dolgov, B.B. Safoklov, D.B. Sukhanov, *Adaptation of a student of a higher educational institution to the educational process in the conditions of distance learning*, in Collection of the 9th International Scientific and Practical Internet Conference "Modern Challenges and Actual Problems of Science, Education and Production", **1**, 476 (2020)

[5] S. Sengupta, *Possibilities and Challenges of Online Education in India During the COVID-19 Pandemic*, International Journal of Web-Based Learning and Teaching Technologies, **17(4)** (2022), DOI: 10.4018/IJWLTT.285567

[6] E.L. Bolotova, A.N. Rizhov, G.M. Kojaspirova, *Pedagogy of higher education* (Prospekt, Moscow, 2021)

[7] C.W. Dereso, K.C. Meher, A.A. Shobe, *COVID-19 Pandemic and Strategizing the Higher Education Policies of Public Universities of Ethiopia*, International Journal of Sociotechnology and Knowledge Development, **14(2)** (2022), DOI: 10.4018/IJSKD.2022040101

[8] A. Parmar, A. Singh, *An Evaluation of Online Education Efficacy and the Measures to Improve It From the Perspectives of Management Students in the Wake of COVID-19*, International Journal of e-Collaboration, **18(1)** (2022), DOI: 10.4018/IJeC.290300

The analysis of foreign countries experience in information technologies application in higher education

Iuliia Kotliarenko[1,*] and *Alvina* Kolesnichenko[1]

[1]Rostov State Transport University, 2, Rostovskogo Strelkovogo Polka Narodnogo Opolchenia sq., Rostov-on-Don, 344038, Russia

Abstract. This paper analyses the use and effectiveness of modern information technologies application in higher education in different countries. This form of learning needs to provide the level of knowledge and competences corresponding to the traditional learning. This article shows advantages and disadvantages of ICT usage in higher education. This article gives the relevance of the comprehensive definition of ICT and their impact on e-learning and teaching processes. The analysis of the pedagogical experience of states with a high level of technological development is interesting and relevant for educational system of Russia This paper focuses on the different models of distance education and the aim of the article is to analyse the present status of the use of digital technologies in foreign countries.

1 Introduction

In the last few years there has been a growing interest in electronic technologies. They quickly penetrated the educational process, completely changing the presentation of the teacher's communication with the student. Having gone from using a printing machine in classes to creating unified information educational spaces, digital, electronic and information and communication technologies (ICT) have become an indispensable element of the modern educational process, opening the whole world to its subjects, filling the content of education with modern meanings and challenges. Such form as distance teaching and learning spread over the world making possible education for all throughout the life. Now there are various models of distance education, its structure and management, different types of its technical and methodological support [1]. An increasing number of scientists believe that the theoretical idea of continuing education should form the basis for the renewal of education systems. Quite recently, considerable attention has been paid to the functions of the teacher. Moreover, in UNESCO's view, encouraging "digital literacy" and ensuring greater mastery of the new information and communication technologies, which should be seen both as educational discipline and as pedagogical tools capable of enhancing the effectiveness of educational services [1]. Undoubtedly, open and distance education increase not only the general educational level of society, but also contribute to the training and retraining of specialists in connection with the transition to new technologies. The teacher's role changes from the main source of knowledge to the moderator, tutor, and consultant. Therefore, the analysis of the pedagogical experience of states with a high level of technological development is interesting and relevant for educational system of Russia.

*e-mail: kotlakot@rambler.ru

In scientific articles of researchers from different countries we can find various terms defining usage of information technologies in higher education. Among them, they are "e-learning", "information and communication technologies", "digital technologies", "computer technologies", "electronic technologies". From our point of view we consider all these terms equal and substitutable [2–4]. Digital technologies are actively applied in distance educational process as well as blended education [5].

In recent years, research on the issue of distance learning has become very popular. In the studies of Ronny Scherer, Sarah I. Hofer and other foreign authors an attempt was made to identify the competence and e-learning perspectives in higher education institutions [6–8]. According to a number of foreign researchers (Michael Sailer, Julia Murböck, Frank Fischer, Albina Bilyalova) and the Russian scientist T.E. Isaeva [12], the process of digitalization of the educational process embraced all countries [9–12].

The relevance of our study is stipulated by the fact of the lack of comparative analysis of the experience of foreign countries in information technologies application in higher education.

2 Methods and materials

Nowadays in Russia we can observe an active process of using information technologies in the educational sphere including foreign language teaching. Numerous digital resources play a significant role in improving an educational process as they attract learners with an interactive form, great opportunities of their application with usage of Internet and multimedia in synchronous, asynchronous as well as blended learning. At same time, information technologies contribute to better assimilation of knowledge and acquired skills and abilities. However, along with advantages of digitalization of higher education we can take notice some difficulties of this process connected with a low rate of its development in Russia in comparison with other countries.

The study is devoted to investigating the experience of applying information technologies in higher education including in the process of teaching a foreign language abroad. According to the Global Competitiveness Report conducted in 2017, the data of competitiveness among 137 countries were given, where Russia ranked 38th place, while Switzerland was on 1st place, the USA was followed, Singapore, the Netherlands, Germany and many others [13]. Regarding the indices of IMD World Digital Competitiveness Ranking 2020 considering digital competitiveness of countries, Russia ranked 43th out of 63 [14].

The leading countries mentioned in these reports have greatly succeeded in such branches as IT technologies, economy, industry, science, etc. as well as in education and their success in all spheres is directly dependent on the high level of education, particularly higher education providing highly qualified specialists. In this article we will try to prove this idea of interaction between digital development of education and world recognition, competitiveness of these countries. The analysis of experience of the countries advanced in this sphere facilitates understanding of the importance of new technologies implementation in the educational process and reveals some gaps of information technologies application in our country and the ways of improving the present situation.

To achieve this goal, the following objectives must be achieved:

- to describe and analyse foreign countries experience in information technologies application in higher education;
- to reveal advantages and disadvantages of using digital technologies in the educational process;
- to find out interaction between digital literacy and competitiveness of the developed countries.

As the main methods of the research, the following are used:

- a method of content analysis, as a result of which an analysis of foreign countries experience in ICT application in higher education was made in order to assess the effectiveness of its use in e-learning teaching;
- systematization and classification;
- empirical research methods—description and comparison—made it possible to analyse digital technologies, to distinguish their advantages and disadvantages.

The theoretical basis of the study was the scientific work of scientists: Ansuman Sar, Satya Narayan Misra [17], T.E. Isaeva [12], Yu. Zhao, Ana Maria Pinto Llorente, Maria Cruz Sánchez Gómez [7], Yuk Ming Tang, Pen Chung Chen et al. [15], Ronny Scherer, Sarah K. Howard et al. [8], Martin Daumiller et al. [16–18] on methodological issues of the use of digital technologies in the process of teaching foreign languages at the university.

3 Results

Nowadays distance education has a strong development potential all over the world. While only 2.4% of the 222 million students around the world have access to international mobility, an opening of the offer of training distance learning on learners residing beyond national borders seems all right to overcome, not only the difficulties, but also the lack of access to the teaching of audiences who do not have the opportunity to study abroad. It is also an asset for the internationalization of educational institutions superior: showcase of the quality of training offered, distance education can be seen as an instrument of promotion abroad. This fact finds evidence in the analysis of experience of using digital technologies in higher education in various countries given below.

3.1 The USA

The United States of America was the first country that developed and introduced computer technologies in the educational process. It began to use the first operating computers ENIAC for educational purposes in 1946 in the University of Pennsylvania. Since then the process of information technologies implementation into learning has constantly been developed and in 2000-s the most of American universities used Internet technologies in education including distant form. Many universities created their own online courses for attracting students from all over the world, for example, Stanford University, University of Michigan, etc. Thus, Massive Open Online Courses (MOOC) appeared and they offer their learners curriculums of various higher educational establishments. The most popular among them—"Coursera" developed by Stanford University (up to 400 courses), "EDX"—a project of Harvard University and Massachusetts Institute of Technology (targeted at mastering a range of disciplines by students of different countries).

The rapid age of the computer technology development affected language education in the USA. It became possible due to wide spread Internet, electronic education, creation of online courses, usage of various digital resources both in practical classes in classrooms and at home remotely. It had an impact on transition of the role of a teacher from a leader to a tutor and changing the way of a language learning by students.

The term CALL (Computer-Assisted language Learning) appeared in pedagogical science in early 1960-s and it is defined as the process of a language teaching by means of computers. First CALL represented a number of repetitive training exercises for forming language skills. Then in 1970–1990-s the emphasis was placed to communicative skills with using speech models, situations and as a result speaking a fluent language. In 1990–2000-s a language teaching technologies included multimedia, Internet applications and the main purpose of CALL was to develop learners' different skills—listening, reading, speaking, writing oriented on the specific sphere of application of acquired language skills and speech abilities in social interaction (the so-called integrative stage). The intelligent stage of CALL (ICALL) lasts from 2000-s until present days and it is distinguished by a high level of artificial intelligence development, using its technologies. Effective self-learning systems are considered the great achievement of the intelligent stage as they allow students to master a foreign language independently in their own regime when it is convenient for every individual and such form of education positively affects the development of abilities to self-study, independency, and motivation to learning a language. ICALL implies using a great variety of intellectual programs, multimedia means and other digital resources aimed at satisfaction of learners' needs in obtaining necessary information, learning authentic materials, forming the required competences. ICALL can be used by means of various devices in different learning formats—synchronously and asynchronously, with distance and blended learning.

In recent years the trend to using mobile versions of educational language programs in the form of applications has greatly increased in the USA as well as in the world. Such technologies got a title MALL (Mobile-Assisted language Learning). This fact reveals understanding the necessity of learners to master a foreign language and apply individual digital courses for achieving this purpose.

3.2 Western Europe

Distance education in Europe was intensively developed in the early 1970s. This was due to the creation of a number of open universities (universities of distance education). Currently, each European country has a group of educational institutions implementing remote programs. The methods of such training are quite well worked out. Training programs using new information technologies, including satellite television, computer networks, multimedia are of great interest. The European Commission adopted the main provisions of the Current Plan for Digital Education in Brussels on January 17, 2018, calculated until 2022. Education and learning as part of continuing education play an important role in creating a European identity based on shared cultural values. Education is designed to help young people formulate and express their opinions, be involved in social processes and thus shape the future of Europe, characterized by developed democracy, solidarity and inclusion. Digital competence is part of the revised European reference system of key lifelong learning competencies that all citizens must possess. Digital competence means the confident and critical use of digital technology and includes the knowledge, skills and attitudes required by all citizens in a rapidly developing digital society. The European System of Digital Competence for Citizens describes digital competence in five areas:

- information literacy and data management skills;
- communication and collaboration through digital technologies;
- creation of modern digital content;
- security in a broad sense—for personal data, for human health and for environmental conservation;
- identify and solve conceptual problems in the digital environment.

One example is the French National Center for Distance Learning (CNED) that was founded in 1939. The scope of its activities is evidenced by the following data: the number of training courses is 2.5 thousand, the number of users is 185 thousand, branches in

120 countries of the world, 5 thousand teachers participate in the development of training courses and educational activities. There are CNED branches in Toulouse and Vanva, Rouen, Rennes, Poitiers, Lyon, Lille and Grenoble. The Distance Education Center offers training in various institutions to everyone. People of almost any age can take part in the program—from the kindergarten to the post-graduate school. In addition, CNED is constantly organizing advanced training courses. The main areas of work of the center are also characterized by the main target groups—"School", "University", "Competition". The latter is attended by people who want to take vacancies in the public civil service. There are also directions "Professional", "Culture" and so on. According to statistics, more than 300,000 people from around the world receive training at CNED every year. Technical tools used in organizing the work of CNED include satellite television, video and audio cassettes, e-mail, Internet, as well as traditional literary sources. If distance learning systems do not use the feedback principle, then the information needed to conduct lectures, seminars and other types of classes is usually centrally recorded on a video cassette or video disc. Further, these materials are sent, including through computer networks, directly to educational institutions where they are used in training sessions. Educational institutions can use these resources through the Internet. Training programs are regularly broadcast on the national television. New Information Technologies used to provide primarily mechanical and automation, electronics and computer science, economy and management training programs, as well as classes in humanitarian disciplines. For example, linguists work with a program that includes information from the Larousse encyclopedic dictionary.

It is interesting to note the experience of the Universidad National de Educacion a Distancia (UNED) in Spain, which recently celebrated its 30th anniversary. It is one of the largest educational institutions in the country. It includes 58 training centers in Spain and 9 abroad (Bonn, Brussels, Geneva, London, Paris, etc.). Its students also have the opportunity to study in New York and Rome. Programs combine the benefits of American and European training systems. Practical training mostly focus on case studies. The formation of the UNED concept was influenced by the Open University of Great Britain, established some years earlier. UNESCO International Chair was organized at the University of Distance Education in 1997 to actively promote research and development in the field of remote education. At the same time, due to the specifics of distance education, the university uses advanced technologies to provide all students with high-quality and uninterrupted services. Participation of UNED in the international academic mobility programs provides the mutual movement of students from one university to another (both in Spain and abroad), with the aim of teaching: ERASMUS, Ibeoamerica Scholarship Prof, PILAR, MOONLIGHT.

3.3 Australia

Australia began to use multimedia in a distant form of education in 1970 and actively implemented open and flexible education with using Internet and modern digital technologies in middle of 1980-s up to present days. The usage of these technologies are aimed at more active interaction between a teacher and a separate student, a teacher and a group of students and communication within a group.

Combination of blended education and the modern digital technologies eliminates borders between traditional and distant forms of education. All universities in Australia provide higher education distantly. This form of training is very popular among students (about 29% who choose distant form), especially among undergraduates and postgraduates who have the formed skills of self-study and research activity. Online lectures strategies for the development of distance education are widespread in Australian universities because of students' workload, a great number of foreign students without good knowledge of English.

One of the most important priorities of Australian government in the sphere of higher education is implementation of distant education in higher schools. All higher educational establishments of the country fulfil national strategies for the development of distance education and meet all necessary requirements aimed at increasing their competitiveness in comparison with other universities of Australia and foreign countries. The system of higher education comprises state and private universities as well as "Non-University Higher Educational Providers". All of them trend to attract more students by ensuring accessible, flexible education based on the modern pedagogical experience including the usage of digital technologies. There are even "Universities of Third Age" representing free or cheap online courses intended for elder people.

All higher schools provide online educational services on the international market; they are especially popular among Asian countries. In order to be required and take the place among the best in the world Australian universities create various national and international associations of distant education, "Open Universities Australia", they exchange their experience in using digital technologies.

The association "Open Learning Australia" was founded in 1992 on the basis of Monash University with the purpose of giving higher education to most of learners of different age and status with various qualification, skills and abilities, for those who couldn't receive a diploma before. It helped applications to choose the appropriate university. In 2004 this company was renamed into "Open Universities Australia" and the association includes 20 main universities. It uses a smart platform "Smartthinking" offering free round the clock training and consulting services, interaction with a teacher, an assessment system. The educational process including teaching a foreign language is carried out by means of digital technologies.

Australian universities faced competition in providing educational services from the USA, European and Asian countries. This fact led to the development of national training platform "Open2Study" offering free online courses competitive to such popular American platforms as "Coursera" and "EDX". "Open2Study" represents interactive weekly modules the assimilation of which requires four weeks and passing a test after each module. "Open Universities Australia" also develop educational platforms for higher schools in Australia, New Zealand, and Great Britain targeted at giving students professional training.

One of the effective ways of promoting Australian universities in the international educational space has become implication of online courses (Massive Open Online Courses) so popular in the USA. Universities of Australia provide their own courses; some of them are placed on global platform including "Coursera", "Open2Study" of "Open Universities of Australia".

Introduction of new information technologies in the educational process changes traditional teaching form in higher schools in Australia and thereby increases a number of higher educational providers and toughens competition among them. The process of transition to distant education leads to lack of information about applicants and it has a negative impact on full-time education. That's why the Australian universities will make efforts to preserve the advantages of full-time education in the terms of digitalization of higher education as well as to give an open access to their courses and to develop new technologies of pedagogical interaction in information environment.

3.4 East and Southeast Asia

A great success in the sphere of higher education has been achieved in the Asian countries. Nowadays Asia takes the leading position in the provision of educational services. And this fact is proved by the high rating of Asian universities according to the world university rating conducted in 2017. The list of the best 200 higher educational establishments in the world included 13 universities from China, 9—from Japan, 16—from South Korea. In addition to these countries, the list comprises high schools in Taiwan, India, Malaysia and Singapore. Twenty-four Asian universities were mentioned as the best among the world ones. This fact finds its explanation in the correct choice of direction for transforming and organizing the whole educational process including higher education in the countries of East and Southeast Asia.

The great improvements in education were made due to effective adoption of the best European and American training models and contribution of their own national values in that process. Therefore, they managed to combine global trends in education with their culture, national traditions and customs formed under the influence of historical, political and socio-economic factors. These countries provided significant funding for the education sector, created international programs for partnership with the leading educational centres of the world. They try to attract foreign specialists to their universities including English teachers chosen from native speakers. High quality education in Asia is achieved due to increasing motivation of students to study and educators—to professional growth and introduction of modern digital technologies.

Information technologies or E-learning in higher schools are widely used because it is a prosperous form of a future specialist's preparation to his professional activity and it gives him an opportunity to be competitive in labour market. E-learning systems represent advanced multimedia technologies incorporating both visual information system and dynamic one that have an impact on learners' sight and hearing and contribute to a better assimilation of the material. E-learning is aimed to solve the problem of "information inequality" in South Korea and it finds application in distant form. In 2009–2010 in the Summit of South East Asian Nations Association (ASEAN) the project of creation of Cyber-university unifying all Asian countries was developed and adopted. It combined the best 27 universities of Burma, Vietnam, Cambodia, Laos, Thailand, South Korea as members of Association of South East Asian Nations. The purpose of this collaboration is to make higher education of these countries both high qualified and accessible by means of digital technologies.

Singapore takes the leading position in the world in digitalization of the educational process at schools as well as at universities. They began their programs of developing information and computer technologies in 1997 and have done well since that time. Both students and teachers have digital competences. All conditions are created for using computer technologies as means of teaching, flexible networking environment made it possible to experiment and choose the appropriate variant of modern technologies in a teaching process. Singapore actively implements various models of partnerships between educational institutions and potential employers. Among them it is necessary to mention national industrial enterprises as well as international IT companies such as "Microsoft", "Lenovo", etc. Flexibility, variability and integrity are the main components of digital educational space.

South Korea is also the leader by introduction of web intellectual environment into the educational process. South Korea began its innovative development in the educational sphere by means of digital technologies in 1997. Nowadays E-learning in this country is high developed at the level of a separate industry providing schools, universities with educational content essential for each particular discipline and as a result ensuring educational and administrative services. South Korea managed to be the leading country in the world by using smart-education.

The modern universities of South Korea, China, Singapore and Malaysia consider one of the chief requirements of education modernization introducing English-learning courses. Among foreign languages, English is highly demanded. It is explained by the fact that English presents a means of global communication and academic mobility. Universities of different countries invite foreign professors for reading courses; students can get a degree in the

Table 1. Using information technologies in higher education of foreign countries

The country	The period of introducing computer technologies in education	Modern state	Advantages	Disadvantages
The USA	1947—the first operating computers, 2000-s—active usage of Internet technologies	Development of ICALL: using mobile versions of educational language programs in the form of applications	Students study independently in their own regime and to develop abilities to self-study	Students' attention is scattered, they cannot focus on the main material
Western Europe, France	1980-s—using Internet and modern technologies	Personalize the entire period of study and reduce the time of communication between teachers and students (CNED)	Flexibility of the training process. Each student take a distance learning course at an individual pace	Distance course helps to have high self-organization skills
The UK	1963—"Air University" in the UK, 1969—the Open University (mail, TV and radio broadcasting, telephone consultations)	The world leaders in the field of open and distance learning, offering its services in the UK, but also in a number of other European countries	A great selection of classes from general education requirements to highly specialized courses in challenging subjects	University studies are largely theoretical and rarely offer much real-world experience
Germany	1974—using modern technologies in the Fern Universität in Hagen	Means of communication between students and teachers are constantly researched, evaluated, improved	Control through virtual consultations, conferences. Students are provided with access to the university library on-line	Teachers and students should have permanent means of communication and access to the Internet
Australia	1970—using multimedia in a distant form of education	Creating online courses, educational platforms	Competitive on the international market of educational services	Accent is particularly made on distant form of higher education
Asian countries	1997—innovative development by means of digital technologies	E-learning is at the level of a separate industry and as a means of academic mobility and global communication	Higher education is required by students all over the world	A local English model combines the norms of international English and national features

English language in a foreign country. International incorporation has a positive impact on the education development. Japan, Malaysia and China attract foreign students and teachers as well as provide language courses and schools in Singapore, Malaysia, Thailand, Bali. Bilingual teachers are also in great demand in Asian countries.

One of the features in English teaching in these countries is that British English is not cultivated as a standard of a "right" language, but different dialects or variants are created and supported by teachers. Such approach leads to students' acquaintance with diversified English variants that provides language awareness as the English language performs the function of nations unification. Variants of English differ from each other by cultural features, peculiarities of mentality. Thus, their own model of the English language (local English model) is developed by bilingual teachers combining the norms of international English and national features. In Singapore bilingual education includes teaching of all disciplines in English and simultaneously learning an official native language. This approach combines students' development of Asian cultural traditions and it leads to forming language competencies necessary for forming a global outlook. It gives Singapore graduates an advantage in competition on labour market.

Taking into account computer technologies application in teaching a foreign language (English or any other, for example, Russian so popular in China) in Asian countries, it is necessary to note that they became widespread in higher school. Students can develop their communicative skills in the sphere of a foreign language by means of using modern social platforms, chats and sites. A great choice of interactive courses enables students to practice a language and acquire necessary skills.

Therefore, the usage of digital technologies in a language education is widely spread in all countries of East and Southeast Asia as well as all over the world. The Asian countries created universities of the world level forming disciplinary and interdisciplinary programs, improving campuses infrastructure and introducing into international cooperation. So great achievements were made in most of Asian countries due to the attitude to education as a great value for the whole culture from the side of all participants of the educational process.

To clearly demonstrate the world trends in using information technologies in the educational process of higher schools, we need to present the research data in the form of a table 1.

The results in the table reveal that information technologies are widely used in higher education all over the world. The process of digitalization of the educational space embraced all countries. Some of them have succeeded in this, others try to keep up and catch up with the leaders, learn from their experience. However, it is evident that each country follows its own way combining its cultural traditions, peculiarities of nation's mentality with the best foreign practices in applying digital technologies in higher schools.

4 Discussion

The conducted research has confirmed our hypothesis about interaction between success of the countries in digital competitiveness ranking the first places (according to the World Digital Competitiveness Ranking) and the level of digital development in the educational sphere. Undisputed leaders in this sphere are the developed countries of Western Europe and the USA that have applied computer technologies for a long time and have a great experience in using them in the educational space including higher schools.

The experience of these countries was adopted by many other countries such as Australia, East and Southeast Asian countries being analysed in the conducted research. Both Australian and Asian universities faced competition in providing educational services from the USA and European countries. This fact explains the necessity of creating their own open courses, educational platforms, etc. in Australia and Asia. A big leap in introducing modern digital technologies in higher education was made by Asian countries which have successfully combined the best European and American training models and contribution of their own national values in that process. Moreover, nowadays Asian universities are in the top of the world best universities.

The conducted study reveals that great achievements in the sphere of digital higher education were made due to the following main factors:

- accessibility of higher education due to introduction of digital technologies and transition to distant or blended forms of education;
- flexibility of higher education because of individual approach to every student giving him an opportunity to study in their own regime when it is convenient for him through using digital resources; and such form of education positively affects the development of abilities to self-study, independency and research activity;
- integrity of higher education implying unity of educational standards, assessment criteria, presentation of general requirements for the formed competencies of a specialist;
- awareness of higher education as a value that contributes to increasing motivation of students to study and educators—to professional growth and mastering digital competencies necessary for modern digital educational space;
- trend to international cooperation with foreign countries that leads to be competitive on the world market.

We also revealed that in all analysed countries foreign language teaching is conducted by means of digital technologies. The most of these countries create educational content themselves providing their own courses of a foreign language and educational platforms, services (the USA, Great Britain, France, Australia) and make them available in open access. Asian countries consider English as a means of global communication and academic mobility and make great efforts in mastering this language inviting bilingual teachers, providing language internships, conducting lectures and practical classes in English as well as using a great variety of information resources both in traditional way on classes at the university or remotely at home.

5 Conclusion

As a result of the given research, we should conclude that all analyzed countries that actively apply digital technologies in higher education including foreign language teaching observe both advantages and disadvantages of digitalization process in educational sphere. First of all, this process requires possession of digital competence of teachers and students. Second, educators observe the drawback of excessive using information resources that is manifested in the fact that students' attention is scattered, they cannot concentrate on the main material, as they are distracted by additional functions, application capabilities, etc. Young people stop reading books, scientific articles that leads to lack of critical and analytical thinking, ability to research activity. Thirdly, excessive use of computer, mobile and other digital means leads to health problems, blurred vision, absent-mindedness, disorder of cognitive abilities. The above-mentioned factors may cause decline in the quality of education, deterioration of results in mastering a foreign language.

In spite of the existing problems, the conducted research proves the effectiveness of information technologies application in foreign language teaching as well as in higher education in general. The successful experience of foreign countries confirms this fact. It is important for Russian education to take the best experience of the educational systems of the leading countries and implement it in higher education of our country. It will ensure the competitiveness of Russian higher education and the whole country as well.

The conducted research has a great practical importance as it contains the detailed analysis of practical usage of information technologies in foreign language teaching and higher education as a whole. It can be applied by foreign language teachers, methodologists and educational authorities for considering foreign experience and further development of its basic ideas as well as implementation of some relevant trends into the Russian higher education system. This is the prospect of further research.

References

[1] *UNESCO Universal declaration on cultural diversity*, Adopted by the 31st Session of the General Conference of UNESCO, 2 November 2001, Paris, **15** (2001), URL `https://www.un.org/en/events/culturaldiversityday/pdf/127160m.pdf`

[2] S.I. Hofer, N. Nistor, C. Scheibenzuber, *Online teaching and learning in higher education: Lessons learned in crisis situations*, Comput. Hum. Behav., **121**, 106789 (2021), DOI: 10.1016/j.chb.2021.106789

[3] M. Fernández-Gutiérrez, G. Gimenez, J. Calero, *Is the use of ICT in education leading to higher student outcomes? Analysis from the Spanish Autonomous Communities*, Comput. Educ., **157**, 103969 (2020), DOI: 10.1016/j.compedu.2020.103969

[4] K.M. Moser, T. Wei, D. Brenner, *Remote teaching during COVID-19: Implications from a national survey of language educators*, System, **97**, 102431 (2021), DOI: 10.1016/j.system.2020.102431

[5] B. Bruggeman, J. Tondeur, S. Vanslambrouck, *Experts speaking: Crucial teacher attributes for implementing blended learning in higher education*, Internet High. Educ., **48**, 100772 (2021), DOI: 10.1016/j.iheduc.2020.100772

[6] V. Milićević, N. Denić, A. Jovanović, *E-learning perspectives in higher education institutions*, Technol. Forecast. Soc. Change, **166**, 120618 (2021), DOI: 10.1016/j.techfore.2021.120618

[7] Y. Zhao, A.M.P. Llorente, M.C.S. Gómez, *Digital competence in higher education research: A systematic literature review*, Comput. Educ., **168**, 104212 (2021), DOI: 10.1016/j.compedu.2021.104212

[8] R. Scherer, S.K. Howard, F. Siddiq, *Profiling teachers' readiness for online teaching and learning in higher education: Who's ready?*, Comput. Hum. Behav., **118**, 106675 (2021), DOI: 10.1016/j.chb.2020.106675

[9] Y. Perbawaningsih, *Plus Minus of ICT Usage in Higher Education Students*, Procedia Soc. Behav. Sci., **103**, 717 (2013), DOI: 10.1016/j.sbspro.2013.10.392

[10] M. Sailer, J. Murböck, F. Fischer, *Digital learning in school: What does it take beyond digital tecnology?* Teach. Teach. Educ., **103**, 103346 (2021), DOI: 10.1016/j.tate.2021.103346

[11] A. Bilyalova, *ICT in Teaching a Foreign Language in High School Procedia*, Procedia Soc. Behav. Sci., **237**, 175 (2017), DOI: 10.1016/j.sbspro.2017.02.060

[12] T. Isaeva, N. Malishevskaya, L. Cherkasova, A. Kolesnichenko, *University faculty motivation to students' distant evaluation*, E3S Web Conf., **210**, 18071 (2020), DOI: 10.1051/e3sconf/202021018071

[13] *QS World University Rankings (2016–2017)*, URL `https://www.topuniversities.com/university-rankings/world-university-rankings/2016`

[14] *IMD World Digital Competitiveness Ranking 2020*, URL `https://www.imd.org/news/updates/IMD-2020-World-Competitiveness-Ranking-revealed`

[15] Y.M. Tang, P.C. Chen, G.T.S. Ho, *Comparative analysis of Student's live online learning readiness during the coronavirus (COVID-19) pandemic in the higher education sector*, Comput. Educ., **168**, 104211 (2021), DOI: 10.1016/j.compedu.2021.104211

[16] M. Daumiller, R. Rinas, J. Hein,, S. Janke, O. Dickhäuser, M. Dresel, *Shifting from face-to-face to online teaching during COVID-19: The role of university faculty achievement goals for attitudes towards this sudden change, and their relevance for burnout/engagement and student evaluations of teaching quality*, Comput. Hum. Behav., **118**, 106677 (2021), DOI: 10.1016/j.chb.2020.106677

[17] A. Sar, S.N. Misra, *A study on policies and implementation of information and communication technology (ICT) in educational systems*, Materials Today: Proceedings (2020), DOI: 10.1016/j.matpr.2020.10.507

[18] J. Calvert, R. Abadia, *Impact of immersing university and high school students in educational linear narratives using virtual reality technology*, Comput. Educ., **159**, 104005 (2020), DOI: 10.1016/j.compedu.2020.104005

Student character building activities as a component of higher education primary educational program

Dmitriy Kozorez[1,*], *Elena* Dolgova[2], *Anna* Korneenkova[2], *Alena* Rumakina[2], and *Chuang* Shi[3]

[1]Department of Navigation and Control Systems, Moscow Aviation Institute (National Research University), 4, Volokolamskoe shosse, Moscow, 125993, Russia
[2]Department of Educational and Methodological Support of Educational Activity, Moscow Aviation Institute (National Research University), 4, Volokolamskoe shosse, Moscow, 125993, Russia
[3]Harbin Institute of Technology, 92, West Da-Zhi st., Harbin, Heilongjiang, 150001, China

Abstract. The paper provides a solution to the task of implementing student's character building activities as a component of the main educational program of higher education in a technical university through the common universal competences formation, the educational process is considered as a single focused process of education and training in the interests of a person, family, society and the state, where education is an activity aimed at personal development, creating conditions for self-determination and socialization of the student.

The processes taking place in the political, economic, and spiritual life of our country and modern society make us look differently at the problems of education in general and at the formation of a harmoniously developed personality, in particular, through the prism of higher education. The most important purpose of higher education in Russia is to train highly moral, intellectually developed, and creatively working specialists, who will become professionals in their field. The higher education currently faces new, more complex challenges. The education should not be limited solely to the transfer of knowledge and retraining of people. It should change a person's attitude to the surrounding social and cultural environment, ensure the suitability of a person to work in rapidly changing labor and production conditions, and contribute to the formation of flexible thinking and attitudes towards dialogue and cooperation.

The Law "On Education in the Russian Federation" interprets education as a unified purposeful process of character building and training in the interests of a person, family, society, and state, while character building is viewed as an activity aimed at personal development and the creation of conditions for self-determination and socialization of students. Character building and training are two sides of a single personality formation process. While training is based on the cognitive activity of a person, providing him or her with a system of scientific knowledge and skills, education is based on the formation of a person's attitude towards life. Usually, the character building of students was more associated with their extracurricular activities, and the upbringing potential of the content of professional disciplines was not given due attention. Today, this issue has been resolved [1]. In 2020, certain amendments [2] were introduced into the Federal Law "On Education in the Russian Federation", which stated that the significance of the character building component was equal to that of the training one at such educational levels as bachelor's and specialist's degrees. Any technical university has its own specificity due to the fact that individuals with previously formed personalities come

*e-mail: kozorezda@mai.ru

to receive higher education. Constituting a top priority in the higher education system, the character building activities are an organic component of pedagogical work, integrated into the general process of training and development of future specialists [3].

Nowadays, it is especially important to ensure the quality of vocational education and training of competitive, mobile, and highly demanded specialists for the labor market. A competitive specialist is a person who satisfies the needs of the market in terms of his or her professional, psychological, moral, and other qualities, possesses the adaptability and mobility, the ability to quickly adjust to the changing conditions, make decisions and take responsibility for them, as well as independently acquire new knowledge and develop skills for mastering new domains.

In accordance with paragraph 2 of Article 2 of the Federal Law "On Education in the Russian Federation", character building is an activity aimed at developing personalities, creating conditions for self-determination and socialization of students on the basis of socio-cultural, spiritual, and moral values, as well as behavior rules and norms, adopted in the Russian society, for the benefit of a person, family, society, and state, fostering among students a sense of patriotism, civic consciousness, respect for the memory of the defenders of the Motherland and the feats of the Heroes of the Fatherland, law and order, a person of labor and the older generations, mutual respect, and caring attitude to the cultural heritage and traditions of the multinational population of the Russian Federation, nature, and the environment. Therefore, the following aspects can be named as the goals of the character building activities for the students of a technical university:

- developing the mindset and the world outlook, as well as updating the system of basic personal values;
- communicating the universal human norms of morality, national foundations, and academic traditions to the students;
- fostering respect for the law, the norms of collective life, the development of civil and social responsibility;
- fostering a positive attitude to work, socially significant dedication and responsibility in business relationships;
- ensuring the personality development of individuals and their socio-psychological support, the formation of personal qualities necessary for effective professional activities;
- identifying and supporting the talented part of students, developing organizational skills, creative potential, and involving students in the processes of their self-development and self-realization;
- forming the culture and ethics of professional communication;
- generating the own demand of individuals for a healthy lifestyle, a responsible attitude to the natural and socio-cultural environment;
- raising the level of safe behavior culture;
- developing personal qualities and attitudes, social skills, and managerial abilities.

Education is a means of character development and the formation of an individual's culture. During the learning process, the students develop views, a scientific worldview, an understanding of the laws of nature, society and thinking, moral and aesthetic ideas, and the ability to follow the norms of behavior in the society and comply with the laws adopted in it. In addition, personality needs, activity motives, social behavior patterns, values and value orientations are also formed.

Character building within higher education is, on the one hand, the assimilation of knowledge and generalized experience of people contained in the disciplines being taught, and, on the other hand, the management of the student's independent work. The training organizes and stimulates the cognitive process, contributing to the formation of the student's attitude towards life and worldview. When confirmed by personal practice, the acquired knowledge, views and ideas become the beliefs of students, their persistent opinions, allowing them to make independent decisions and choose the desired line of behavior in specific life situations, including those arising during professional activities. The character building impact of academic disciplines manifests itself in three main areas: the formation of favorable personality traits, the creation of a positive attitude towards the disciplines being studied, and the development of interest to the chosen training area or specialty. All studied disciplines, including humanitarian and socio-economic, mathematical and natural science, general professional, specialized disciplines of a profile, and elective disciplines, are designed to contribute not only to the professional, but, above all, to the spiritual and moral development of future specialists. The character building potential of the management of students' independent work is focused on the formation of their independent thinking skills, their own work styles, self-discipline and willingness to work hard, responsibility, self-control, and self-esteem.

The humanitarian and socio-economic disciplines are almost the same for all technical areas of training and specialties. They are aimed at forming a holistic worldview and desire to solve socio-political problems among future graduates.

The purpose of mathematical, natural science and general professional disciplines is to provide the future graduates with fundamental knowledge in the field of their professional activities, which will allow them to navigate within the framework of their professions, ensure their mobility and ability, if necessary, requalify for another direction of training or specialty, master new equipment and technologies.

An integrated approach towards teaching, combined with character building activities, requires conveying to the students the ideas about the integrity of the world, as well as the social purpose and humanistic content of any technical and scientific achievements. A future specialist should have the understanding of the need for harmonious coordination of science, technology, and people.

Character building activities should be carried out even when higher education institutions operate in an online format, and they should provide for the involvement of the students themselves. Education should include not only knowledge and skills, but also spiritual, moral values that shape the personality and unite the society.

Since the character bringing activities for a university student are focused on the purposeful formation of the civil and social image of a future professional and specialist and on the development of patriotic qualities of a person, capable of a flexible, evaluative thinking process, building a constructive dialogue and fruitful relationships in various fields, then their contents are embodied in the following areas:

- Civic and patriotic education, focused on the development of a set of general civil values and legal culture through the involvement in social and civic activities, the development of a sense of indifference to the fate of the Fatherland, to its past, present and future for the purpose of motivating students to fulfill and defend the interests of the Motherland.
- Legal education, focused on the formation of students' legal awareness and legal culture, as well as the development of students' stable orientation towards law-abiding behavior.
- Spiritual and moral education, focused on the development of the value-semantic sphere and spiritual culture, moral feelings and a strong moral foundation, moral and aesthetic personality qualities of the students, as well as on fostering the culture of interethnic communication.
- Cultural and creative education, which implies familiarization with the material and non-material objects of human culture and provides a humanistic orientation for the activities of the student youth.

- Sports and recreational education, focused on the formation of a healthy and safe lifestyle culture, the development of the ability to preserve and strengthen one's health, the formation of such personality qualities as high morality, aesthetic sense, positive moral, collectivist, vocational and physical qualities, as well as on ensuring moral, psychological, and physical readiness for the professional activity.
- Environmental education, focused on the development of environmental awareness and sustainable environmentally-friendly behavior.
- Professional and labor education, focused on the development of psychological readiness for professional activities within the chosen occupational area.
- Scientific and research education, focused on the formation of research skills and critical thinking, as well as motivation for scientific and research activities.

The requirements for employees, who are graduates of technical specialties, are currently changing. Employers increasingly value specialists, who are inclined to be creative, since routine operations are more and more automated. Thus, the professions, which will be in demand in the future, will be associated with creativity and the ability to consider problems comprehensively, instead of looking at them from the point of view of a narrowly focused specialist. Therefore, all the aforementioned areas of activities are applied as a comprehensive set. They cannot exist separately. All of them are targeted at the formation of a personality within one's professional sphere and ensuring the completeness of individual's moral development.

Nowadays, higher education provides for the application of a competence-based approach, which allows to determine the content of character building and the quality of training of a specialist, reveals an integrated characteristic of the graduate training quality, which is the result of education and a set of interrelated personality traits [5].

Most Russian universities shall ensure the compliance of their educational process with the requirements of the Federal State Educational Standard of Higher Education (hereinafter referred to as the Educational Standard). One of the Educational Standard requirements to a university graduate is the development of a set of universal competencies (hereinafter referred to as UC), which are common for all training areas/specialties [6]. For the inclusion of character building activities into the primary educational programs of higher education (hereinafter referred to as the Primary Educational Programs), a higher education institution needs to incorporate character building work programs and a calendar plan of the relevant student activities with the appropriate forms of their certification [4]. For the bachelors' and specialists' areas of training, it is advisable to implement character building activities as part of the formation of general universal competencies, defined by the categories of health preservation, intercultural and interethnic communication, patriotism, and citizenship. As part of the Primary Educational Programs, the character building work program is implemented through the reflection of the directions of character building activities within academic disciplines. Thus, a university provides specialists who are not only brought up in a patriotic and cultural sense, but are also able to comprehensively solve the professional tasks assigned to them, taking into account legal, social, and economic norms and restrictions. The correlation of the directions of character building activities with academic disciplines is presented in table 1, which uses a bachelor's degree program as an example.

For each discipline, educational outcomes (knowledge, skills, abilities) are determined, which are correlated with indicators for determining the formation of competences related to the character building activities of a student. An example of the correlation between learning outcomes associated with character building activities of a student and indicators for determining the formation of competence according to UC-1 of the bachelor's degree program is presented in table 2. This ensures monitoring of the quality of the character building component implementation within the student educational process by checking the level of knowledge, skills, and abilities acquired by him or her.

Table 1. Correlation of the directions of character building activities with academic disciplines, using a bachelor's program as an example

Direction of character building activities	Code and name of the universal competence from the Educational Standard	Discipline
Scientific and educational, Civic and patriotic	UC-1. Capable of performing data search, critical analysis and synthesis, applying a systematic approach to solving the assigned tasks	History, Philosophy, Introduction into Aviation and Aerospace Technology
Legal, Professional and labor, Scientific and educational	UC-2. Capable of determining a range of tasks within the framework of the assigned goal and choose the best ways to solve them, based on the current legal norms, available resources, and restrictions	Legal Science, Economic Theory, Fundamentals of Management, Economics of the Industry
Spiritual and moral	UC-3. Capable of carrying out social interaction and fulfilling his or her role within a team	Sociology, Fundamentals of Psychology
Professional and labor, Scientific and educational	UC-4. Capable of carrying out business communication in oral and written forms in the state language of the Russian Federation and a foreign language(-s)	Foreign Language
Civic and patriotic, Spiritual and moral, Cultural and creative	UC-5. Capable of perceiving the intercultural diversity of the society in its socio-historical, ethical, and philosophical contexts	History, Philosophy, Sociology, Cultural Studies
Professional and labor	UC-6. Capable of managing his or her time, building and implementing a self-development path based on the principles of education throughout the entire life	Fundamentals of Psychology
Sports and recreational, Environmental	UC-7. Capable of maintaining a proper level of physical fitness to support full-fledged social and professional activities	Physical Training (sports sections), Physical Training, Ecology
Sports and recreational, Environmental	UC-8. Capable of creating and maintaining safe living conditions in everyday life and during professional activity, intended to preserve the natural environment, ensure sustainable development of the society, including in the event of a threat of emergencies and military conflicts or their actual occurrence	Health and Safety Training

Table 1. (Continue)

Direction of character building activities	Code and name of the universal competence from the Educational Standard	Discipline
Professional and labor, Spiritual and moral	UC-9. Capable of applying basic development pathology knowledge in social and professional spheres	Fundamentals of Psychology
Professional and labor	UC-10. Capable of making informed economic decisions in various areas of life	Economics of the Industry, Economic Theory, Fundamentals of Management
Legal	UC-11. Capable of forming an intolerant attitude towards corrupt behavior	Legal Science

Table 2. Correlation between learning outcomes (knowledge, skills, abilities), associated with character building activities of a student, and indicators for determining the formation of competence, using the bachelor's degree program UC-1 competence as an example

Competence code and name	Indicator code and name	Discipline	Learning outcomes (knowledge, skills, abilities)
UC-1. Capable of performing data search, critical analysis and synthesis, applying a systematic approach to solving the assigned tasks	UC-1.1. Searches and critically evaluates the information needed to solve a problem	History	Possess the tools of historical knowledge and analysis in order to form a civic position
		Philosophy	Be capable of applying the terminology, categorical apparatus used in philosophy when arranging problem solving based on the principles of systematic thinking
			Have an idea about the main scientific paradigms and research programs
	UC-1.2. Proposes and evaluates various options for solving the problem based on the application of a systematic approach	Philosophy	Know the scientific image of the world, features of the modern civilization development, and basic concepts of the global philosophy

Table 2. (Continue)

Competence code and name	Indicator code and name	Discipline	Learning outcomes (knowledge, skills, abilities)
UC-1. Capable of performing data search, critical analysis and synthesis, applying a systematic approach to solving the assigned tasks	UC-1.2. Proposes and evaluates various options for solving the problem based on the application of a systematic approach	Philosophy	Be capable of distinguishing between a disciplinary and an interdisciplinary approach to the study of objects as systems and the focus of the study towards identifying ties within an object and between an object and the environment
	UC-1.3. Formulates a professional activity problem statement based on the critical analysis of the available data and on the systematic approach	Philosophy	Be capable of selecting, understanding, and structuring modern scientific literature, identifying its theoretical and methodological directions
	UC-1.4. Uses the knowledge of the main paths for developing and enhancing the aviation and aerospace equipment during the professional activities	Introduction into Aviation and Aerospace Technology	Know the basics of the design of aircraft and their systems, the main directions of the modern development of science and technology in the field of aviation and aerospace equipment

The proposed approach to the implementation of the student character building activities as a component of the primary educational program is applicable both for Russian students and for foreign students who are trained at the universities of their countries during their junior student years and at a Russian university during their senior student and receive two diplomas of higher education at the same time, Russian and foreign. For example, for students from the People's Republic of China, spending their first student years at a university of their country, the civil and patriotic education is composed of studying the history and philosophy of China.

The calendar plan of the student's character building activities and their forms of certification is a sequence of mastering the mentioned disciplines in accordance with the educational schedule, indicating the form of certification for each discipline. To implement character building activities within the teaching process, it is advisable for the professor to be guided by the following principles:

- strengthening the connection between learning and life, theory and practice;
- establishing interdisciplinary ties within the disciplines of various cycles, taking into account the profession being mastered;

- involving students in the social activities of production teams and forming their professional mentality;
- studying the advanced experience, including foreign data.

Group cooperation, team rivalry, brainstorming, business games, organizational self-administration, and many other forms are those interaction processes, which foster creativity, freedom, tolerance, citizenship, and other qualities of the students when their arrangement is well thought out.

Therefore, for the purpose of implementing paragraphs 1 and 2 of Article 12.1 of the Federal Law No. 273-FZ dated December 29, 2012 "On Education in the Russian Federation" [4], the Moscow Aviation Institute has introduced the character building activities of students into all its Primary Educational Programs starting from September 1, 2021. The introduction of the proposed model of student character building activities into the Moscow Aviation Institute educational process provides an opportunity to enhance the process of training in technical specialties by developing creative individuals, who will be able to work in a team, self-educate, apply a comprehensive approach while solving professional problems, make decisions and take responsibility for them. The educational process with the implemented character building activities is the future of training personnel for high-tech areas of the Russian industry.

References

[1] P.N. Osipov, *Character building activities in an innovative higher education institution* (BRONTO, Kazan, 2019), 264 p.

[2] Federal Law No. 304 dated July 31, 2020 "On the amendments to the Federal Law 'On Education in the Russian Federation' (in terms of student character building)"

[3] E.O. Mikhailova, S.P. Kulikov, S.V. Novikov, *On the issue of implementing Character building activities in higher education institutions*, Modern Pedagogical Education, **12**, 37 (2019)

[4] Federal Law No. 273-FZ dated December 29, 2012 "On Education in the Russian Federation"

[5] I.V. Rudenko, L.V. Alieva, U.A. Kustov et al., *Theoretical foundations of the formation of general professional competencies of students within the character building process of educational institutions* (Publishing House "Academy of Natural History", Moscow, 2017)

[6] D.A. Kozorez, E.A. Dolgova, A.V. Korneenkova, A.V. Rumakina et al., *Aerospace education in Russia. Support of educational programs* (MAI Publishing House, Moscow, 2020), chapter 3.2

Providing personnel for enterprises of high-tech industries of the Russian Federation in the field of artificial intelligence

Dmitry Kozorez[1], *Yuri* Sledkov[1], *Vladimir* Busurin[1], *Pavel* Kudryavtsev[1], *Nadezhda* Makarenkova[1,*], and *Ting* Dong[2]

[1]Moscow Aviation Institute (National Research University), 4, Volokolamskoe shosse, Moscow, 125993, Russia
[2]Shanghai Jiao Tong University, 800 Dongchuan RD. Minhang District, Shanghai, China

Abstract. The concept of artificial intelligence from the point of view of regulatory documents is revealed, the main areas of development of this technology are given. The current needs of the industry in highly qualified personnel in the field of artificial intelligence are considered. Technical problems are presented that can be solved with the help of artificial intelligence systems, in particular, with the use of computer vision. A generalized structure of a control system for a moving object using artificial intelligence technologies and a conditional diagram of data processing processes are given on the example of a mobile system that moves along a given trajectory. The main approaches to the training of highly qualified personnel in the field of artificial intelligence systems in leading Russian and foreign universities, as well as in the framework of specialized educational courses, are given. The connection of training in the field of artificial intelligence systems with other applied disciplines is established.

1 Introduction. The concept of an artificial intelligence system

To control complex technical objects with the required accuracy and reliability, especially in conditions of uncertainty in the external environment, a promising solution is the use of artificial intelligence technologies, which can increase the reliability of the information output signal, improve the quality of control and the speed of decisions, eliminate the possibility of emergencies due to human factor [1–3]. According to GOST R 59277-2020 "Artificial Intelligence Systems. Classification of artificial intelligence systems" any automatic system that functions without human intervention, and an automated (hybrid) system, the functioning of which is carried out under human control, is an artificial intelligence system (AIS) [4]. At the same time, there are no recommendations for separating automatica systems (with feedback, adaptive, with identification, etc.) from learners and self-learning systems. The classification procedure, the requirements for the quality of functioning, the accuracy of training, the volume of the training samples are determined only by the developer.

The importance of developing intelligent control systems that would solve a wide range of functional tasks, including an independent search for solutions under various external conditions, is emphasized, among other things, by the Decree of the President of the Russian Federation "On the Development of Artificial Intelligence in the Russian Federation", which

*e-mail: n.a.makarenkova@yandex.ru

 https://doi.org/10.1051/shsconf/202213701011

states the need to develop technologies AIS in relation to various areas of knowledge and its use in the development of modern technical systems and control systems for various kinds of objects [5].

The National Strategy for the Development of Artificial Intelligence for the Period up to 2030 includes [5]:

- support for scientific research;
- development and evolution of software;
- increasing the availability and quality of data;
- increasing the availability of hardware;
- creation of an integrated system of regulation of public relations;
- increasing the level of provision of the Russian market of artificial intelligence technologies with qualified personnel.

At the same time, the last point can be noted as the most important, since the introduction of AIS in solving the priority tasks of the industry can be carried out only with the involvement of highly qualified specialists in the relevant field.

2 Industry Requests and Application of AIS

Let's consider some urgent problems, the request for the solution of which comes from high-tech industries and involves the use of AIS technologies, on the example of vision systems for moving objects.

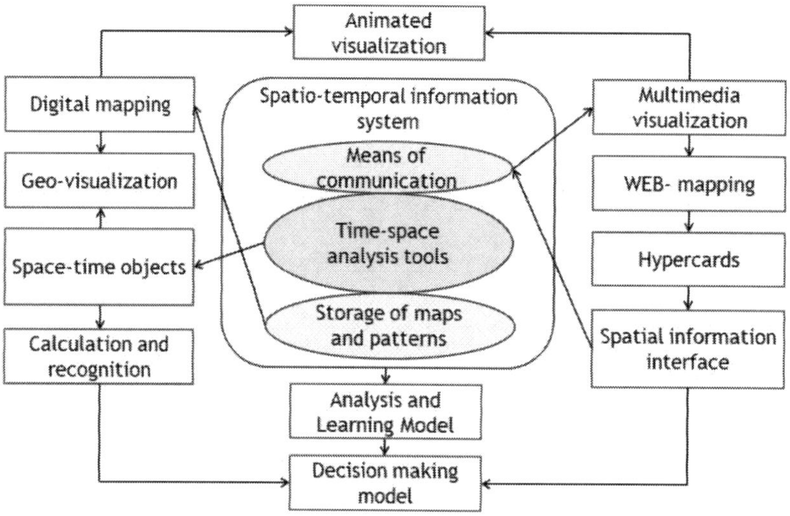

Figure 1. Generalized structure of a control system of a moving object using artificial intelligence technologies

When creating vision systems, it is necessary to solve the following algorithmic problems [2, 6–9]:

- preliminary image processing from optoelectronic sensors (changing the brightness range, highlighting the boundaries of objects, forming related fragments, etc.) in order to improve the vision systems of a technical object;

- development of algorithms for geometric alignment of images from video sensors and synthesized images in real time;
- development of algorithms for detecting and recognizing reference zones (points, lines, contours, etc.), as well as foreign objects located on the trajectory of movement, etc.

In the figure 1 shows a generalized structure of a control system of a moving object using artificial intelligence technologies.

Thus, in order to develop visual recommendations, the vision system must be able to solve the following tasks [7–10]:

- visually identify and recognize the terrain and obstacles;
- to carry out mutual referencing and integration of sensory and geospatial information;
- to carry out the synthesis of the image based on the integration of navigation information and information from the databases of the terrain and obstacles;
- to carry out photogrammetric processing of 2D and 3D data to create digital terrain models;
- to obtain an estimate of the distance from a moving object to a selected point;
- to obtain estimates of the trajectory and angular position of the moving object in space, as well as to form comparative estimates of the position of the moving object relative to a given route;
- to form the values of control signals for the timely correction of the trajectory and angular motion of the moving object;
- to carry out human-machine communication using interfaces.

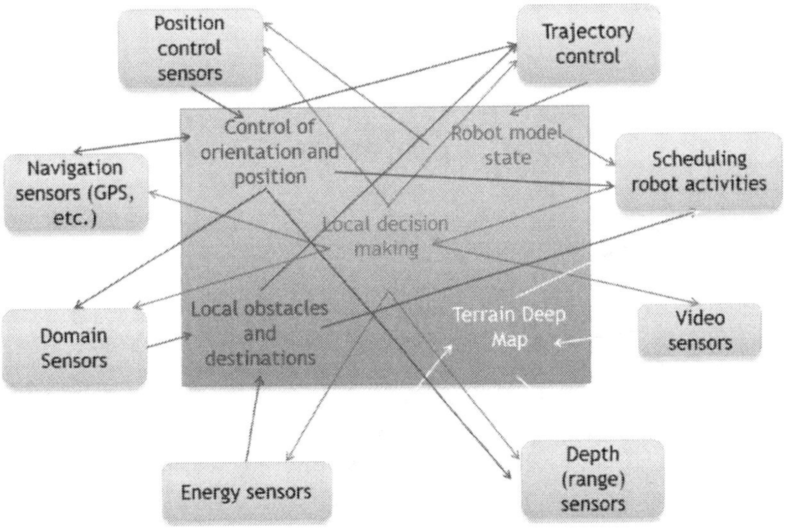

Figure 2. Schematic diagram of data processing processes in a mobile system for simultaneous position localization and terrain mapping

Simultaneous mapping of the surrounding area and assessment of the position of the robot allows, for example, to avoid collisions with obstacles or entering a danger zone. Scanning laser rangefinders (SLRs) and a visual odometry system are most commonly used as hardware for such systems. SLRs scans a sector of the surrounding space and returns a vector of distances to obstacles, for example, every degree.

To perform mapping and localization tasks, a mobile object must be equipped with an appropriate set of sensors, as well as a different set of computing facilities and software modules for data processing [2, 11]. Sensors, in fact, are operational data sources, and the corresponding buffers for the accumulation of sensory information are operational information databases [10]. In the figure 2 shows a schematic diagram of data processing processes in a mobile system for simultaneous position localization and terrain mapping.

The data from the corresponding sensors are sent to the units for controlling the orientation and position of the moving object, assessing its condition, finding local obstacles and destinations, compiling a depth map of the area, and, ultimately, in the unit for local decision making.

In the figure figure 3 shows the architecture of a typical system that implements the basic methods of visual odometry and terrain mapping.

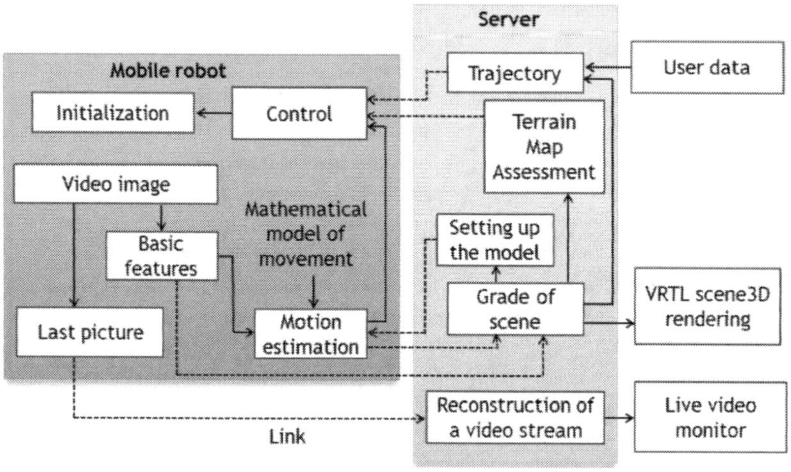

Figure 3. The architecture of a typical system that implements the basic methods of visual odometry and terrain mapping

First, preliminary processing of data on the current state of a moving technical object (for example, from a gyroscopic unit) and visual information is carried out [2]. Then one part of the data is transmitted either via a communication channel for subsequent processing (for example, to solve the tasks of planning the subsequent movement and identifying the desired objects in the search zone) to an external server, or processed by a decision-making unit, provided that it can be placed on a mobile technical object. Another part of the processed data is directly used on site to stabilize the position of a moving technical object.

3 Training of personnel in the field of AIS in the universities of the world and the Russian Federation

In accordance with the Decree of the President of the Russian Federation on the development of artificial intelligence, the directions of increasing the level of provision of the Russian market of artificial intelligence technologies with qualified personnel include [5]:

1) development and implementation of educational modules within educational programs of all levels of education, advanced training and professional retraining programs;

https://doi.org/10.1051/shsconf/202213701011

2) attracting organizations operating in the field of artificial intelligence to participate in activities aimed at the development of general and professional education;

3) an increase in the number and an increase in the attractiveness of competitions and olympiads aimed at developing the intellectual and creative abilities of students.

From the classification list of GOST R 59277-2020 "Artificial Intelligence Systems. Classification of artificial intelligence systems" it follows that a specialist in AIS should have a more extensive amount of knowledge, in comparison with a specialist in conventional automatic control systems [4]. Convergent knowledge, provided, among other things, through the integration of mathematics, natural science, and social and humanitarian education, is gaining priority. The relationship of training in the field of AIS with other applied disciplines is shown in figure 4.

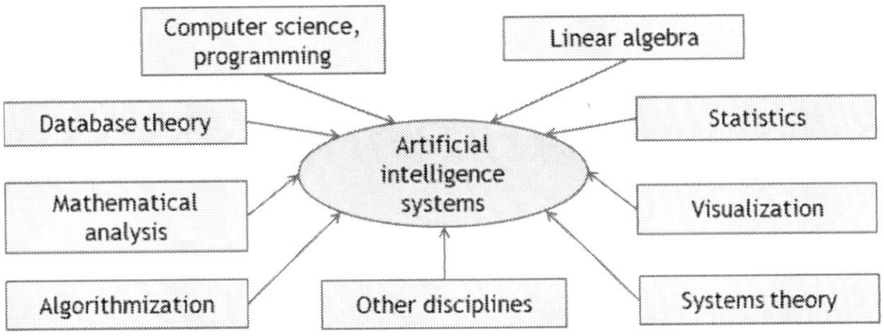

Figure 4. The relationship of training in the field of AIS with other applied disciplines

Almost all the leading technical universities in the world declare the possibility of training specialists in AIS. Moreover, artificial intelligence is being poured into completely different areas of training—business informatics, economics, information systems and technologies, applied informatics, mining, management, ecology and nature management, biology, control in technical systems, automation of technological processes and many others. At the same time, universities provide students with the opportunity to gain competencies in the most relevant areas related to artificial intelligence: computer technology, cryptography, intellectual analysis, parallel and distributed data processing, mathematical modeling, digital hardware, robotics, machine learning, and others. This makes it possible to train specialists capable of creating a technological base for ultra-fast computing, quantum computers, complex communication systems, software systems for machine learning, analyzing bigdata, predicting and calculating risks, modeling complex processes. At the same time, an important feature of the educational process is in-depth mathematical training of students, with an emphasis on the use of mathematical disciplines in the field of information technology.

However, the intensive development of AIS technologies has led to the fact that many universities are even so unable to quickly change their curricula and prepare the appropriate teaching staff for the implementation of new achievements of artificial intelligence. In this regard, a lot of commercial courses have appeared, mainly on-line, giving the necessary competencies to specialists of generally accepted profiles. Among such programs, there are a lot of courses for people who do not have the required foundation of knowledge, and only want to understand the essence of the new directions of AIS. However, in a number of cases, together with universities for which it is difficult to quickly change the curriculum, such courses allow to produce highly qualified specialists in the field of AIS. The duration of training on such courses, with appropriate software and technical support, is from 7 to 15 months.

4 Conclusions

In GOST R 59277-2020 "Artificial Intelligence Systems. Classification of artificial intelligence systems" there are no recommendations on the separation of automatic systems from learning and self-learning systems. However, an AIS specialist requires a broader body of knowledge and technical horizons than an automatic control specialist. So, more in-depth knowledge of the theory of probability and mathematical statistics, the theory of random processes, the theory of automatic systems, the theory of algorithms, mathematical logic, the main sections of the theory of computer vision and others is needed. The urgent task now is to create a comprehensive classification of methods and tools used in the creation of AIS and training in the field of AIS.

There are real practical tasks in various fields of activity that can be solved using artificial intelligence technologies (automatic visual route control, mapping of the surrounding space, estimation of the position of the robot, integration of sensory and geospatial information, etc.). In this regard, most of the leading universities in the world and in Russia offer educational programs that include the development of competencies in various aspects of artificial intelligence in a wide range of areas.

Mastering a complex body of knowledge in the field of artificial intelligence using training programs provides highly qualified specialists who are able to make sophisticated technological decisions to imitate human cognitive functions (including self-learning, finding solutions without a predetermined algorithm and achieving insight).

References

[1] Minimum Aviation System Performance Standard (MASPS) for Enhanced Vision Systems (EVS), Synthetic Vision Systems (SVS), Combine Vision Systems (CVS) and Enhanced Flight Vision Systems (EFVS), DO-315B (RTCA, 2011)

[2] V. Busurin, S. Zheltov, P. Kudryavtsev, *Aerospace Visual Space-Time Systems* (MAI Publishing House, Moscow, 2015)

[3] P. Harrington, *Machine Learning in Action* (Manning Publications, New York, 2012)

[4] GOST R 59277-2020 "Artificial intelligence systems. Classification of artificial intelligence systems"

[5] Decree of the President of the Russian Federation No. 490 of October 10, 2020 on the development of artificial intelligence

[6] R. Gibb, R. Gray, L. Scharff, *Aviation Visual Perception: Research, Misperception and Mishaps* (Ashgate, Burlington, 2010)

[7] S. Zheltov, O. Vygolov, Y. Vizil'ter, *Aviation Systems for Enhanced and Synthesized Out-of-Cockpit Vision*, Polyot. All-Russian Scientific-Technical Journal, **1**, 33 (2013)

[8] Enhanced Flight Vision Systems, AC 90-106A (FAA, 2013)

[9] Kollsman General Aviation Vision System—GAViS (last checked: 2016-01-01), URL `http:/www.kolsman.com`

[10] Y. Vizilter, *Enhanced and Synthetic Vision System Implementation based on IMA Platform*, in International conference "State and development prospects of integrated modular avionics", 29–30 October 2012, Moscow (2012), 51 p

[11] E. Fedosov, V. Kos'yanchuk, N. Sel'vesyuk, *Integrated Modular Avionics*, Radioelectronic Technologies, **1**, 66 (2015)

Methodological approach to the implementation of a "2+2+2" higher education model at a technical university

Dmitriy Kozorez[1,*], *Iurii* Sledkov[2], *Anna* Korneenkova[2], *Maria* Toporova[3], *Alena* Rumakina[4], *Maria* Pushkareva[5], and *Weixing* Zhou[6]

[1]Department of Navigation and Control Systems, Moscow Aviation Institute (National Research University), 4, Volokolamskoe shosse, Moscow, 125993, Russia
[2]Department of Control Systems, Informatics and Electric Power Engineering, Moscow Aviation Institute (National Research University), 4, Volokolamskoe shosse, Moscow, 125993, Russia
[3]Department of Aviation Technology, Moscow Aviation Institute (National Research University), 4, Volokolamskoe shosse, Moscow, 125993, Russia
[4]Department of Educational and Methodological Support of Educational Activity, Moscow Aviation Institute (National Research University), 4, Volokolamskoe shosse, Moscow, 125993, Russia
[5]Department of Engineering Economics and Humanities, Moscow Aviation Institute (National Research University), 4, Volokolamskoe shosse, Moscow, 125993, Russia
[6]Harbin Institute of Technology, 92, West Da-Zhi st., Harbin, Heilongjiang, 150001, China

Abstract. The paper provides a solution to the task of implementing a "2+2+2" model of higher education while providing the students with an opportunity to choose the direction of training starting from the third year of studies; reviews the process of developing Primary Educational Programs as a complex technical object through the application of a systematic approach; identifies the conditions and restrictions that allow avoiding the destruction of the systematic nature of education, and, therefore, the degradation of its quality; proposes the principles of applying individual educational paths for achieving the target of granting the students with an opportunity to obtain one or more qualifications.

Nowadays, Russian universities operate using a "4 + 2" educational system, which means four years for obtaining a bachelor's degree and two years for getting a master's one. At the same time, there are cases when students, studying in a bachelor's degree program, understand that the chosen specialty does not suit them, dramatically change their area of training within a master's program, and, in fact, get a second profession during it. Some do this many years after they have completed their bachelor's or specialist's studies.

During his address on January 15, 2020, the President of the Russian Federation, set the task to the Federal Assembly of the Russian Federation to provide the students, mastering the primary educational programs of higher education (hereinafter referred to as the Primary Educational Programs), with an opportunity of choosing the direction of training starting from the third year of studies [1].

In Russia, experts have started active development of a new model of higher education, which is called a "2+2+2" model. The new model of higher education assumes the following gradation of the educational process (refer to figure 1):

- During their first 2 years of studies, students will master general subjects included in the range of "general developmental" ones, expanding the horizons and suitable for all special-

*e-mail: kozorezda@mai.ru

ties, such as: history, philosophy, legal science, health and safety training, disciplines of the physical training block, disciplines of the basic economic block, disciplines of the block of project activities, disciplines of the basic foreign language studying block, disciplines of mathematical, physical, natural science and engineering modules, and disciplines of the information technology module.

- The following 2 years will be dedicated to a deeper study of specialized subjects that will be directly related to the specialty, meaning that the choice of a specific profession will be made after completing the first 2 years of studies, during which students will be able to understand which specialty suits them and which subjects and areas of science are more interesting for them.
- Another 2 years of studies will be devoted to a master's program, where certified bachelors can further deepen and refine their knowledge and skills within the framework of an already mastered profession. It is also allowed to take a new direction, but this will require a certain "base", forming minimal required knowledge and understanding.

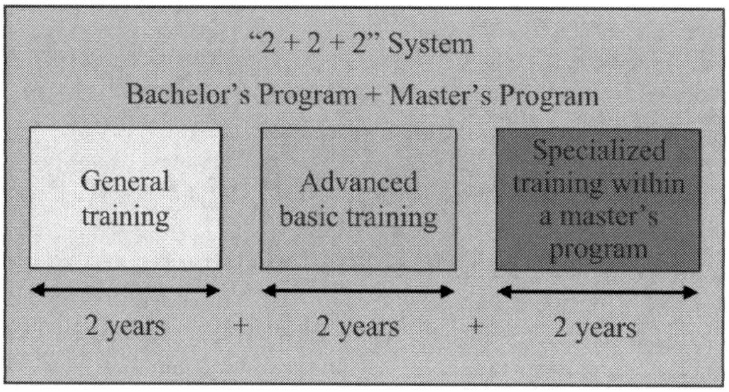

Figure 1. New system of higher education per "2 + 2 + 2" model

The "2 + 2 + 2" model is a more detailed version of the current education system, which is aimed at an accurate and effective outcome. Applicants will submit documents to a specific institute, which means that some of the basic subjects will have a certain "emphasis": economic, mathematical, humanitarian, technical. Then, while studying them, the student will be able to understand which of the areas/specialties is closer to him or her, which area of knowledge is more interesting, easier, etc.

It was the creation of the "2 + 2 + 2" model which was the background for introducing the new rules for admission into universities in Russia starting from 2021: applying to an institute (instead of a specific direction/specialty), enlarged competitive selection, a "postponement" of the selection of a specific profession (till the 2nd–3rd year of studies), etc.

At the same time, the following amendments were introduced into the Federal Law "On Education in the Russian Federation": "Primary educational programs may include competencies, which are attributed to one or more professions, specialties, and areas of training, taking into account the possibility for students to obtain several qualifications simultaneously". Therefore, the second task set for all Russian higher education institutions is to provide an opportunity for a student to obtain two or more qualifications [2].

In order to effectively implement the proposal of the President of the Russian Federation, it is first necessary to understand how the additional professional module will relate to the main part of the curriculum and be introduced into the educational process, taking into account all the restrictions governed by both the requirements of the education system and the requirements of the industry.

For this purpose, let us consider the Primary Educational Program development process as a complex technical object through the application of a systematic approach [3].

In this case, the designing of a complex technical object is usually viewed as a process of creation, transformation and presentation in an acceptable form, carried out for an object intended to perform a certain set of functions. Initially, the image of this object is expressed in the form of certain demands represented in terms of a set of its functional properties and characteristics, which must ultimately be transformed into a description of a technical object in terms of its structure, composition, dimensions, etc.

Let us apply this approach to the development of line Primary Educational Programs. The term "line" is used here in the sense of "widespread Primary Educational Programs, designed to support a significant number of jobs".

In this case, the designing process is carried out iteratively through sequences of stages, terminating with the creation of the next prototype of the design object with the generation of the corresponding intermediate definitions that summarize the solution of certain problems and are used to make the following design decisions.

While technical objects are handled in the design process (during formalization), various representations, reflecting their functional, structural, parametric, and constructive and graphic aspects, usually get tied to the development objects. In the course of designing, the following sequence of transformations is implemented:

$$F \Rightarrow S \Rightarrow P \Rightarrow G,$$

where F is functional presentation, S is structural presentation, P is parametric presentation (reflection of the appearance of the item being designed in the parametric space), and G is structural and graphical presentation.

As applicable to the Primary Educational Program development process, the domain F, functional presentation, contains the qualification characteristics of a graduate (competences); the domain S, structural presentation, includes elements of competence structuring, identification of their specific components, development of the competence formation flows; the domain P, parametric presentation, defines the Primary Educational Program support (methodological support, staffing, resources); and the domain G, structural and graphical presentation, is the development of schedules, curricula, work programs for disciplines (modules).

During the development of technical objects, the cycle of a single complete iteration of research activities consists of the following three main stages:

1. Identification of the demand for new functional qualities based on the efficiency analysis and identification of the deficient functions of the system fleet and elements currently available and being developed (development objects).

2. Generation of alternative tactical and technical ways to eliminate the deficit.

3. Synthesis of the structural and functional image of the alternative concepts in terms of their performance indicators within the framework of variable constraints.

The first stage consists in identifying the need for new functional properties of the designed object. Here, it is required to define the deficiency of the functional properties of the currently available "fleet" of Primary Educational Programs.

The second stage is associated with the generation of alternative technical options and tactical ways to eliminate the properties deficit identified at the first stage. This leads to creating alternative concepts, i.e. a set of basic technical and tactical properties of the system, ensuring the achievement of the specified functions.

The third stage is dedicated to solving the problem of system synthesis, taking into account the existing constraints and the environment in which the development object will operate, applying the criteria of "efficiency vs. expenses vs. development time" type, and evaluating the dynamics of the system potential enhancement.

Based on one of the main provisions of systems analysis, which states that decisions should be made using the largest possible set of alternatives, this stage provides for the consideration of the greatest possible number of options for solving the original problem of eliminating the deficiency of properties through the creation of a new design object, update of the existing items for their better adaptation to the changed tasks and application conditions, and modification of the existing items and entire systems or the ones being developed for incorporating the desired functional qualities into them.

A morphological set of alternative solutions is formed for the main areas of functional properties deficiencies. The main life cycle stages and the resulting requirements, in particular, the ones to the system operation, material and technical support, standardization, as well as the possible development time and the necessary expenses are taken into account.

All the aforementioned aspects are also applicable when designing Primary Educational Programs.

When describing the algorithm for the development of "line" Primary Educational Programs, it is necessary to define the concept of program efficiency, since we are talking about the optimization of educational programs. Program efficiency will be interpreted as the achievement of the required qualification characteristics (competencies) within the time and cost constraints.

Within the framework of the outlined approach, the algorithm for developing "line" Primary Educational Programs consists of two stages.

Stage 1. External (conceptual) design.

The first stage, which is one of the product life cycle upper stages, is the external (conceptual) design, consisting of a sequence of steps as follows:

Step 1.1. Identifying the deficiency in staffing for the industry and the labor market.

The goal setting at this step provides for monitoring the needs of the industry and the labor market.

This step is when it is already possible to draw preliminary conclusions about the number of required jobs.

Step 1.2. Definition of job functions and activities at the workplace.

The goal setting at this step provides for selecting an area of training or specialty, within the framework of which a Primary Educational Program will be created or updated.

At this step, it is possible to determine the areas and spheres of professional activities, professional tasks and objects of professional activity of a graduate. It is also possible to carry out the initial "quality" optimization of the Primary Educational Program. If no suitable Primary Educational Program is available, a decision is made to develop a new direction (specialty), a federal state educational standard for higher education (hereinafter referred to as the Educational Standard).

Step 1.3. Formulation of the Primary Educational Program mastering outcomes—general professional and specialized competencies, determination of the Primary Educational Program scope and competence achievement indicators.

The goal setting at this step provides for defining the direction of the Primary Educational Program.

At this step, the final decision is made on the development of a new direction (specialty) or update of the existing Primary Educational Programs.

The main participants during the first stage process are members of the expert community of employers and university methodologists.

The main data sources are professional standards, profession data sheets, sectoral qualifications frameworks, regulatory documentation of industry excellence centers.

The main conclusion to be made at this design stage is what "consumption rate" the Primary Educational Program being developed is going to have. This shows whether it is necessary to open a new direction of training with significant admission control numbers or the problem can be solved by using individual learning paths of the graduates.

Stage 2. Primary Educational Program detailed design.

The second stage of the "line" Primary Educational Program development algorithm is associated with the preliminary and detailed designing of the product and consists of a sequence of steps as follows:

Step 2.1. Competence structuring—identification of the didactic units.

The goal setting at this step provides for defining educational technologies and making a decision on the curriculum structure.

The competence structuring process is a very important and time-consuming stage in the Primary Educational Program development. During it, it is first necessary to define a set of didactic units that contribute to the formation of the given competence and constitute its essential content. At the same time, we deliberately move away from the discussion and selection of disciplines, instead of which we talk about didactic units as a more stable category of the educational content. The discipline names vary according to the preferences and traditions of a particular university [4].

This step also includes finalizing the competence achievement indicators and determining the level of competence formation and the form of its certification. The levels of formation of didactic units, and, therefore, competencies, are defined in accordance with the levels of mastering the educational material (Bloom–Anderson's Taxonomy) [5].

Step 2.2. Identification of the projected learning outcomes (knowledge, skills, abilities).

The goal setting at this step provides for the optimization of the Primary Educational Program through the assessment of the achievability of the given competence formation level.

This step includes identifying the disciplines (modules) o be incorporated in the syllabus and the sequence of their mastering (discipline flows). It defines the required support for Primary Educational Program: methodological support, staffing, and material resources.

Step 2.3. Development of a calendar training schedule, curriculum, detailed programs of disciplines (modules), detailed practice programs, educational and methodological set, evaluation means bases.

The goal setting at this step provides for generating a completed Primary Educational Program.

Note that, in general, the process of Primary Educational Program detailed designing takes place within the university premises, but when determining the levels of competence formation, it is necessary to involve expert of the employers, since they can provide a more professional evaluation of this parameter, thus significantly influencing the effectiveness of the Primary Educational Program.

An evident trend in higher education today is the customization of training. As a consequence, it becomes necessary to form an individual? learning path (hereinafter referred to as ILP) with active involvement of the student. If we analyze most of the engineering Primary Educational Programs, developed based on the order of the industry and taking into account all the constraints of the educational system (financial, technical, etc.), we can conclude that

https://doi.org/10.1051/shsconf/202213701012

the potential for the implementation of student individual learning paths is quite limited. To expand it, it is proposed to review the approaches providing for deeper unification of Primary Educational Programs during junior years of studies.

Further on, let us define what the unified training during the first two years of studies means: whether it is some common training for all areas, or common training within the framework of an enlarged group of specialties and directions (hereinafter referred to as an Enlarged Group), or common training within a higher education institution.

Most Russian universities shall ensure the compliance of their educational process with the requirements of the Educational Standard. The Educational Standard defines the requirements for a university graduate in the form of a set of competencies: universal competencies (hereinafter referred to as UCs), which are common for all training directions/specialties, and general professional competencies (hereinafter referred to as GPCs), which vary depending on the training direction/specialty [6].

If we assume that unified training is common for all directions, then the first two years of studies in all universities should be composed of the disciplines, forming only the UCs; otherwise, the Educational Standard requirements will not be fulfilled. For technical universities, such an approach is not acceptable.

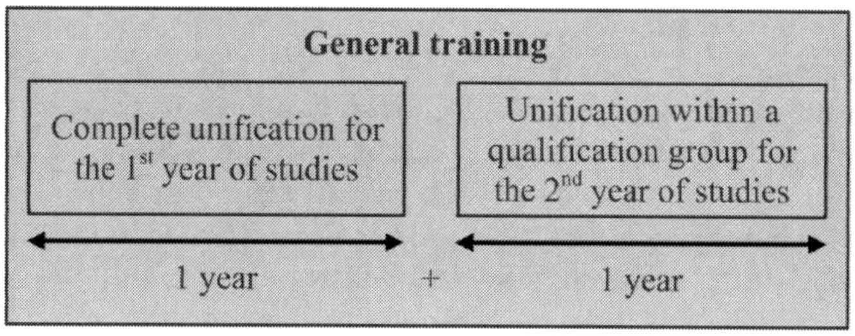

Figure 2. General training—2 years

The following approach is proposed for ensuring common training during the first two years of studies (refer to figure 2):

1) complete unification for the first year of studies;

2) unification within a qualification group for the second year of studies; for this, all the Enlarged Groups implemented at the university should be combined into groups depending on their qualification attribute. The following qualification groups are proposed:

   - **Group 1**—design and technological directions;
   - **Group 2**—design and systems engineering directions;
   - **Group 3**—physics and mathematics directions;
   - **Group 4**—information and communication directions;
   - **Group 5**—service and operational directions;
   - **Group 6**—economic and socio-humanitarian directions.

The complete unification for the first year of studies supports the following:

1. Image of the university—all university students, regardless of the direction they entered, study the same set of disciplines during their first year, except for one or two disciplines necessary to support the direction of training.

2. Student adaptation to the new environment.

3. Familiarization with the areas of professional activities—the “Core 2” module contains a discipline named “Introduction into professional activity”, which provides for the participation of all the graduating departments of the relative directions, representing their training areas and profiles. Thus, a student has the opportunity to choose an enlarged group of directions/specialties during the 2nd semester of the 1st year of studies (refer to figure 3).

4. Mastering of the competencies at the level of “*Has an idea of...*” [5].

The unification of the 2nd year of studies within a qualification group supports the following:

- deliberate selection of the qualification group during the 2nd study semester;
- deliberate selection of the direction of training during the 3rd study semester;
- deliberate selection of the profile during the 4th study semester;
- basic part for the formation of generalized competencies of the qualification group;
- mastering of the competencies at the level of “*Ready to master...*” [5].

A student gets an opportunity to choose the direction/specialty of training during the 3rd semester of the 2nd year of studies, and the profile/specialization—during the 4th semester of the 2nd year of studies (refer to figure 3). The student performs it being assisted by a mentor from the graduating department, who is a specialist in the field of education, supporting students throughout their individual development paths [7].

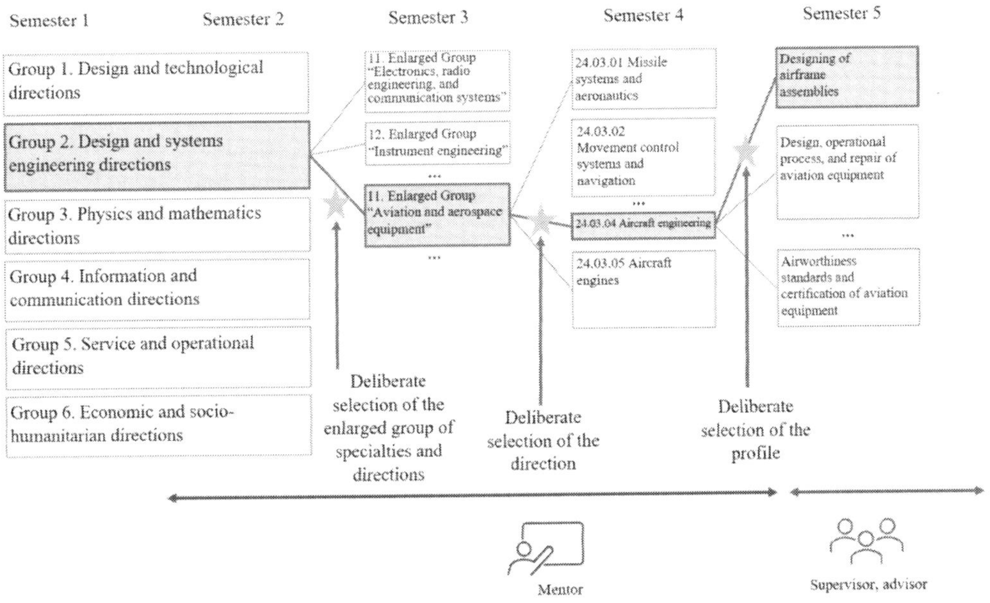

Figure 3. The selection of direction/specialty and profile/specialization

There are several directions within the framework of engineering specialties, and ILPs shall provide the opportunity to obtain a diploma in several qualifications. In Russia, graduates of higher education institutions will be able either to obtain several qualifications within a single engineering specialty or to change the directions of training. The universities will decide on their own, what specialties will be available for combined training. The demand for individual learning paths is growing every year, but such a transition requires profound organizational and administrative transformation within the higher education institution. At the stage of applying to a university, such a concept as the "irreversibility of choice" of the selected profession for students is gradually disappearing. This means that if a student is currently admitted for the first year of studies within a certain direction of training, he or she finishes it, except for some individual cases when students move from one direction to another. Today, this irreversibility of choice is gradually disappearing and eroding, as the students can refine, readjust their learning process in accordance with their desires. The students select disciplines from a set of elective disciplines determined by the demands of the industry. The ILPs affect not only students. The faculty and teaching staff gain additional motivation to create high-quality training content, since now the learners themselves are entitled to decide, the lecture of which professor to attend.

The main potential of the ILP implementation is in the differentiation of the competence formation levels for different qualifications and directions. This provides "time", which can be used to master additional competencies or to deepen further the study of the main ones. The proposed algorithm for the ILP development, based on the needs of the industry, through the formation of competencies, didactic units, and learning outcomes makes it possible to implement the differentiation of the competence formation levels for different qualifications and training directions. Let us consider the stated differentiation using as an example the formation of the "Readiness to apply and develop working technical documentation and ensure its compliance with standards, technical requirements specifications, and other regulatory documents, as well as carry out the documentation of the completed design activities" competence, which is provided through a different set of disciplines with the different content amounts for various qualification groups. For example, for the group of design and technological directions, it is formed by the "Engineering and computer graphics", "Projective geometry", "Technical drawing", "Metrology and standardization", and "Quality systems" disciplines in the amount of 13 credit units, while for the group of design and systems engineering directions, it is formed by the "Engineering and computer graphics" and "Metrology and standardization" disciplines in the amount of 7 credit units. The "Engineering and computer graphics" discipline, which forms the aforementioned competence, has a greater scope for the group of design and technological directions than for the group of design and systems engineering directions. Thus, for the group of design and systems engineering directions, certain "time" is vacated either for in-depth mastering of another core competence of a direction or profile, or for acquiring new additional professional competencies. In a technical university, it is possible to master additional qualifications in related areas with similar basic training.

Advanced basic training during the next two years of studies (refer to figure 1) provides the following:

1. In-depth mastering of professional competencies at the level of "*Can apply in a standard situation...*", "*Possesses the methodology...*", "*Capable of developing...*" [5].

2. Individual learning paths.

3. An opportunity to obtain several qualifications within the training directions of a single qualification group, or within the related directions having similar basic training.

For the implementation of the new higher education model, the following curriculum structure is proposed:

- **"Core 1" module**—a compulsory part to be mastered by all students, regardless of their training direction/specialty (dedicated to the formation of universal competencies);

- **"Core 2" module**—a compulsory part to be mastered by all students within a single classification group (dedicated to the formation of general professional competencies);
- **"Profile foundation" module**—a compulsory part to be mastered by students of a certain profile/specialty (dedicated to the formation of professional competencies);
- **"Elective disciplines" module**—a mobility module supporting ILPs, designed for mastering elective disciplines (dedicated to the formation of additional professional competencies).

The first two years of studies within bachelor's/specialist's programs are composed of the disciplines belonging to "Core 1" and "Core 2" modules.

The disciplines of the "Profile foundation" module start from the third year of studies within bachelor's/specialist's programs.

The "Elective disciplines" module consists of the "General elective disciplines" block and the "Profile/specialization elective disciplines" block. For the "General elective disciplines" block, it is proposed to allocate 15–20 credit units. The introduction of new university-wide elective disciplines should be approved by the commission based on the provided scope of assessment tools for the discipline and student opinions.

For the implementation of the university-wide elective disciplines, it is proposed to introduce the competence formulated as "Capable of applying personality developmental training technologies, providing motivational support and formation of reflexive abilities as the basis for conscious self-governing during professional activities", which should be formed by general elective disciplines with the same indicators and learning outcomes.

The "General elective disciplines" block provides for the incorporation of the following five thematic blocks:

- **Block 1.** Digital development and information society.
- **Block 2.** Personal effectiveness, leadership, and teamwork.
- **Block 3.** Management technologies.
- **Block 4.** Innovative entrepreneurship.
- **Block 5.** Globalization processes and world culture.

The requirements for the university-wide elective disciplines are as follows:

- Absence of pre-requisites.
- All disciplines having 3 credit units, with the selection of one discipline per semester in semesters 3, 4, 5, 6, 7.
- The possibility for a student to choose a discipline of one block or several blocks. The student can choose one elective discipline from each block or "collect" disciplines belonging to 1–2–3–4–5 blocks.
- The same way of assessment—a graded examination or a rating.
- Part of the lecture material can be posted on online resources.
- If necessary, more in-depth study can be arranged through optional disciplines.
- 3 successive stages of the selection: for "A" level students, for "B" level students, and for the remaining ones.

The elective disciplines of the "profile/specialization" are introduced starting from the 5th semester. Selecting the disciplines they are interested in, the students choose their own paths corresponding to the profile/specialization. Thus, a sequence of elective disciplines is constructed, allowing to master the professional competencies of the profile/specialization. The student's decision on which path to take depends on the demands of the industry. Customers from an enterprise send a request to the educational institution, based which a set of competencies is formed and the ILPs are generated. On the other hand, there are dependences in the process of studying the disciplines, which must be observed when a student chooses an ILP. Therefore, while forming their ILPs, the students can be assisted by a mentor from the graduating department, who has the information necessary for this process [7].

The "2+2+2" system will allow to make the direction of training more specific, determine the student's future profession more accurately, and minimize the rate of expulsions from the first two years of studies. Moreover, students have two opportunities to choose a profession: during their bachelor's degree programs and during their master's degree programs. At the same time, this does not exclude the possibility of transferring to another profile, another institute or university.

The new education system has the following advantages:

- more informed choice of the profession;
- more practical training;
- general professional and personal development;
- in-depth study of the specialty during senior years of studies.

This structure has enhanced flexibility and removes one of the barriers for the most promising students. They get the opportunity to design a training program based on their individual needs and preferences.

Thus, a rigid "framework" is replaced by a "modular construction set", which can be assembled by the students on their own, who are certainly supported by experienced supervisors, advisors, and mentors (refer to figure 3). Mentors assist in choosing the ILP. The graduate project supervisors and industry advisors, assigned for on course projects and practices, assist in implementing the ILP through the selection of research and development topics. This is primarily relevant for motivated applicants, who have not yet determined their professional area but would like to adjust the curriculum to their own particularities as much as possible.

The proposed implementation of the new higher education model "2+2+2" at a technical university allows to provide the following:

1. Deliberate choice of the direction/specialty during the 2nd semester of the 1st year of studies and choice of the profile/specialization during the 4th semester of the 2nd year of studies, which and minimizes the rate of expulsions from the first two years of studies.

2. ILPs, focused on the students and the needs of industry and the market, which ensure the flexibility and mobility of higher education, provide differentiation of the competence formation levels for different qualifications and training directions, therefore vacating "time", which can be used either for in-depth mastering of another core competence of the direction or profile, or for acquiring new additional professional competencies. In addition, the ILP implementation makes the university more prestigious for applicants.

3. The set of elective disciplines is formed by the commission, considering student opinions as one of the factors; this determines the rating of the professors, increases the competition between the professors, makes it possible to award the most demanded professors, motivates the students to study, and additionally inspires the teaching staff to create high-quality educational content.

4. The possibility of changing the direction/specialty, which allows to keep the student numbers in the entire university.

5. Adaptation of the higher education to the new requirements of the industry and the market through the creation of a set of elective disciplines, research and development topics.

6. The ability to create a training program for promising students, based on their individual needs and preferences.

7. The ability to adapt the training of line personnel, intended to fill a significant number of jobs, to the new requirements of the industry and the market.

For the specialist's degree programs, the proposed approach to the construction of the higher education system can be implemented in the "2+2+1.5" form, where the first two years of studies are dedicated to general training, the next two years of studies are composed of advanced basic training, and the final 1.5 years of studies are formed by specialized training.

Thus, the following has been performed to ensure the possibility of choosing the direction of training starting from the third year of studies for the students mastering higher education programs, as well as to provide the students with an opportunity to obtain two or more qualifications:

1. The methodology for the unification of the general training during the first two years of studies has been proposed.

2. The conditions and limitations, allowing not to destroy the systemic nature of education, and, therefore, not to degrade its quality, have been identified.

3. The ILP application principles have been proposed.

4. The possibility of obtaining several qualifications has been evaluated.

References

[1] Address of the President of the Russian Federation to the Federal Assembly of the Russian Federation dated January 15, 2020, MM-P13-441

[2] Federal Law No. 144-FZ dated May 26, 2021 "On Amendments to the Federal Law 'On Education in the Russian Federation'' (in terms of students obtaining several qualifications)

[3] D.A. Antonov, R.M. Babich, Yu.P. Balyko, A.M. Zherebin et al., *Aviation of the Russian Air Force and scientific and technical developments. Combat sets and systems of yesterday, today and tomorrow* (Drofa, Moscow, 2005)

[4] *Experience in the implementation of competence-based educational and methodological sets in the educational institutions of vocational education on the example of certain modern areas of science and technology: guidelines* (KSTU Publishing House, Kazan, 2009)

[5] A.L. Galinovsky, E.I. Pudalova, M.I. Toporova et al., *Methodological approaches to the transition from professional standards to educational programs in the aircraft industry*, in III All-Russian Scientific and Practical Conference "Quality Management in Education": notes for the report, 21–22 April 2010, LETI, Saint Petersburg, Russia (2010)

[6] D.A. Kozorez, E.A. Dolgova, A.V. Korneenkova, A.V. Rumakina et al., *Aerospace education in Russia. Support of educational programs* (MAI Publishing House, Moscow, 2020), chapter 3.2

[7] M. Gorchakov, *Knowledge by choice*, Oblako, **3(12)**, 28 (2019)

Experiences in engineering design training at Samara University

Vladimir Kurenkov[1] and *Alexander* Kucherov[1,*]

[1]Samara National Research University, 34, Moskovskoye shosse, Samara, 443086, Russia

Abstract. This paper aims to share experience gained by Samara National Research University in training students in engineering degree programme "Manned and Unmanned Spacecraft and Space Systems". Core items of engineering degree curricula are projects because they basically form practical experience of future design engineers. Main items of the course project "Calculation of main parameters and generation of land remote sensing satellites conceptual design based on regard to required efficiency indices" are discussed. These include acquisition and processing of statistical data on space systems and satellites; determination of satellite orbit parameters; determination of mass-dimensional characteristics of spacecraft on-board systems and construction; formation of on-board systems; spacecraft construction models and assembly model; choice of launcher and spacehead model development. The activity is supported by appropriate training materials and software.

1 Introduction

Engineering design education is based on knowledge accumulated by students in the process of learning science, mathematics and engineering disciplines. In the context of educational programmes in space technology, those engineering disciplines include disciplines pertaining to spacecraft systems, construction and project concept.

Students also study: statement of problems of spacecraft design; choice of basic design parameters; criteria of design solutions; monocriteria and multicriteria optimization; design and construction automation.

Experience in design and engineering also evolve from academic and job training, graduation thesis preparation and course projects activities as a whole. Critical role of the latter activity is emphasized by accreditation requirements of Association for Engineering Education of Russia (AEER) and European Network for Accreditation of Engineering Education (ENAEE) [1, 2].

Curriculum of the engineering degree programme "Manned and Unmanned Spacecraft and Space Systems" implemented in Samara University includes 5 course papers and 5 course projects.

In this paper, experience in realizing a course project in the discipline "Construction and design of spacecraft, boosters and spacestations" is presented. The subject of the project is "Calculation of main parameters and generation of land remote sensing satellites conceptual design based on regard to required efficiency indices".

*e-mail: alexandersk@mail.ru

2 Project statement

The initial data for the design is as follows:

- wavelength range of observation;
- number of wavelength ranges;
- ground resolution;
- swath width;
- surveillance frequency;
- satellites system productivity;
- operational efficiency;
- operating lifetime;
- upper limit of the satellite orbit altitude;
- form of the satellite body section;
- number of solar cell battery panels.

Students are also provided with their work schedule (see table 1). Recommended software, along with industry-standard application, includes the software "EFKAN", which was developed by researches of Samara State Aerospace University (SSAU) and SRC "Progress".

Table 1. The work schedule

No.	Items content	Recommended software	Volume, %	Deadline, weeks
1	Acquisition and processing of statistical data on space systems and satellites	Internet, Microsoft Office, Matlab	10	3
2	Determination of sun synchronous orbit parameters	EFKAN	15	4
3	Determination of electro-optical telescopic system (EOTS) parameters	Mathcad, Excel	30	5
4	Choice of data acquisition and digitizing system, storage device and radio data link	Internet, Microsoft Office	35	6
5	Choice of on-board control system and motion control system	Internet, Microsoft Office	40	7
6	Calculation of gyrodyne complex parameters	Mathcad	45	8
7	Calculation of complex propulsion system parameters	Mathcad	50	9
8	Calculation of temperature control system parameters	Mathcad	55	10

Table 1. (Continue)

No.	Items content	Recommended software	Volume, %	Deadline, weeks
9	Calculation of electric power supply system parameters	Mathcad	60	11
10	Formation of on-board systems, spacecraft construction models and assembly model	SolidWorks	75	12
11	Formation of preliminary weight report	Mathcad, Excel	80	14
12	Choice of launcher and spacehead model development	SolidWorks	95	16
13	Processing of explanatory note and project presentation	Microsoft Word	100	17

3 The project contents

3.1 Acquisition and processing of statistical data on space systems and satellites

This item is necessary, first, for taking into account recent advantages in spacecraft design and, second, for clarification of regression relationships that are to be applied in methodical ware and software in order to provide use of adequate mathematical models (accuracy of the spacecraft mass-dimensional characteristic calculation depends upon accuracy of models). Statistical data must be collected on satellites' characteristics and on their structures.

The data available to a design engineer, is typically heterogeneous. An attempt to develop a model based on the whole data population may result in a rather low quality of the model. Therefore, the data must be previously divided into more or less homogeneous groups. The problem may be solved by using cluster analysis [3].

3.2 Determination of satellite orbit parameters

First, orbit type is chosen. For most current technology land remote sensing satellites (LRSS), sun synchronous orbits are used. Minimum orbit altitude $H_{\min}$ may be expressed as follows:

$$H_{\min} = \frac{R_E}{\sin(\gamma_{\max})} \cdot \sin\left(\frac{L}{2R_E} + \gamma_{\max}\right) - R_E,$$

where R_E is the Earth radius, L is given swath width and $\gamma_{\max}$ is maximum deviation angle of the EOTS line-of-sight from the nadir.

Altitude and inclination of sun synchronous orbit are related as follows:

$$\cos(i) = \frac{\mu p^2(2\pi - T_C\omega_E)}{2\pi\varepsilon N},$$

where i is orbit inclination, $\mu = 398600$ km^3/s^2 is the Earth gravitational parameter, p is orbit focal parameter, $T_C = 86400$ s is mean solar day duration, $\omega_E = 0.729211 \cdot 10^{-4}$ s^{-1} is the

Earth angular velocity, $\varepsilon = 0.003352$ is ellipticity of Earth, N is number of the satellite daily orbit passes.

If surveillance system includes several satellites, it is possible to use single orbits, but ground traces must be shifted relative to each other [4, 5].

Then, students have to estimate surveillance frequency for several survey objects and operational efficiency. Frequency of the same object surveillance depends essentially upon orbit parameters and the object latitude, while operational efficiency (operativeness of video data delivery) depends upon orbit parameters and placement of ground information receiving point. Estimation of surveillance frequency and operational efficiency can be performed with the help of the software "EFKAN". The software uses simulation of the satellite orbital movement and its target turns; as a result, statistical expectations, mean-square deviations, distribution functions and density functions of surveillance frequency and operational efficiency are calculated.

Figure 1 shows the satellite trace and zones of visibility as an example. Boundaries of the satellite zones of visibility are calculated at given time intervals [5].

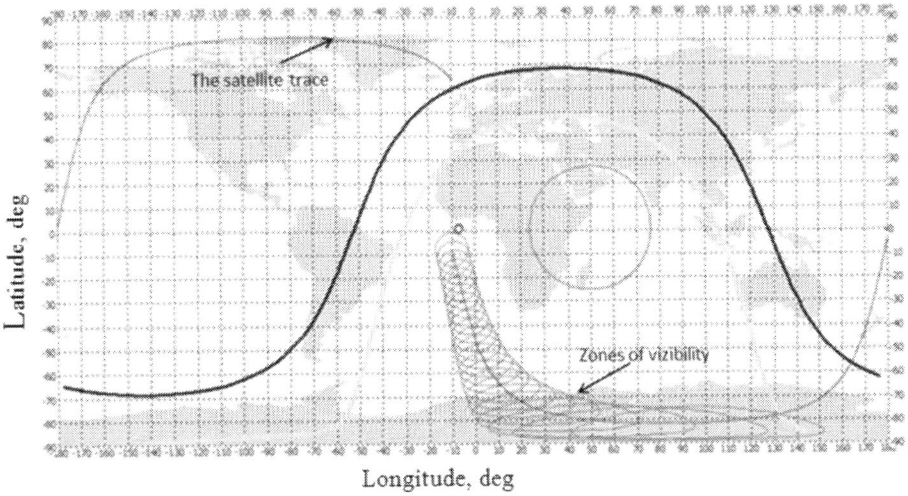

Figure 1. The satellite trace and its zones of visibility

When periodicity and operational efficiency are determined, orbit parameters, number of satellites, number of ground information receiving points are clarified, and necessity of data-relay satellite for providing operational efficiency is analysed.

3.3 Determination of mass-dimensional characteristics of spacecraft on-board systems and construction

If there is no equipment with necessary characteristics, the problem of developing new systems arises. Statistical data on various on-board systems of existing spacecraft or, alternatively, appropriate mathematical models based on physical principles of the systems operation are used. Models for estimation of target systems characteristics must include models of electro-optical telescopic system, data acquisition and digitizing system, storage device, radio data link. Models for estimation of support systems characteristics must include models of on-board control systems (gyrodyne complex, system of gyrodynes desaturation, stellar coordinate measurement unit, etc.); temperature control system (including passive and active components); electric power supply system; complex propulsion system; construction of spacecraft modules and hinged elements; antenna feeder systems; on-board cable system and on-board electronics [6–11].

Such models have been developed by researches of SSAU and SRC “Progress”. They allow determining not only on-board systems and spacecraft masses, but also their dimensions and inertia moments. Those models are introduced in the education process and are accessible to all students.

One example is the model for determining parameters of charge-coupled device (CDD) array. First, it is necessary to choose an optical circuit and to estimate the equipment mass, dimension and power consumption. Choice of an optical circuit is a multicriterion problem and it has no clear solution [5, 11]. Ritchey–Chrétien optical circuit, for the most part, is considered. Cut-off frequency transmitted by a lens ν_c may be revealed,

$$\nu_c = \frac{D}{\lambda f},$$

where D is the lens diameter, λ is the wave length and f is the optical system focal length. Considering that limit resolution (difraction resolution) in focal plane is one-half of cut-off frequency, it equals

$$\Delta L_f^{\text{dif}} = \frac{\lambda f}{2D}.$$

Then, provided that electro-optical telescopic system axis is directed to nadir, ground resolution may be obtained as follows:

$$\Delta L_G = \frac{\lambda H}{2\nu_0 D},$$

where ν_0 is the ratio of operational frequency to cut-off frequency. Thus, minimal diameter of lens required to obtain ground resolution is

$$D_{\text{min}} = \frac{\lambda H}{2\nu_0 \Delta L_M}.$$

Liner dimension of elementary photodetector δ with account for some factors such as Nyquist frequency, characteristics of CCD arrays now in use, internal equipment noise, etc., may be determined as

$$\delta = \frac{\lambda}{(0.81 \pm 0.008)} \cdot \left(\frac{f}{D}\right),$$

where f/D is lens aperture. Then, required focal length may be determined as

$$f = \frac{\delta H}{\Delta L_M}.$$

The CCD array length depends upon required field-of-view of the EOTS.

3.4 Formation of on-board systems, spacecraft construction models and assembly model

First, the satellite structure is formed based on determined mass-dimensional characteristics of spacecraft on-board systems and construction. Then, 3D models of spacecraft elements are formed in one of CAD-systems (normally, SolidWorks system is used).

Formation of the assembly model must be realized with account for set limitations and chosen criteria. As limitations, a design engineer may set areas for location of some elements (apparatus, for instance), areas for location of hinged construction elements, taking account of their potential shielding, and envelope.

Formation of the assembly model may be realized by several methods. We recommend the method of iterative "addition" of on-board systems and construction around EOTS by criterion of minimum mass and minimum inertia moments of spacecraft [4, 12]. The method is used in SRC "Progress" for "manual" iterative cycles of spacecraft design. Design works sequence using this method is shown in figure 2. An example of LRSS assembly model formed by a student is shown in figure 3.

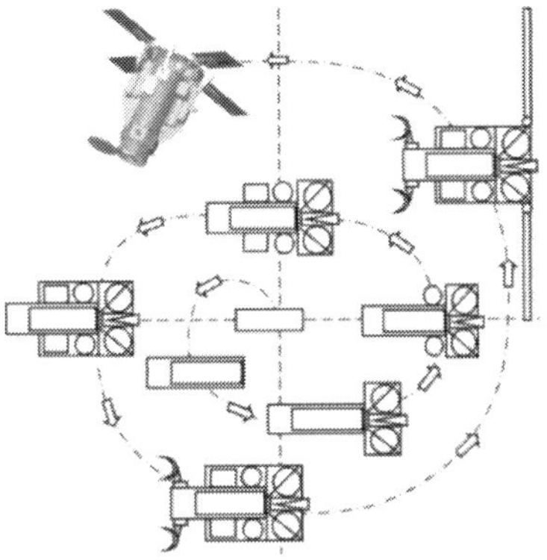

Figure 2. Sequence of spacecraft design concept formation

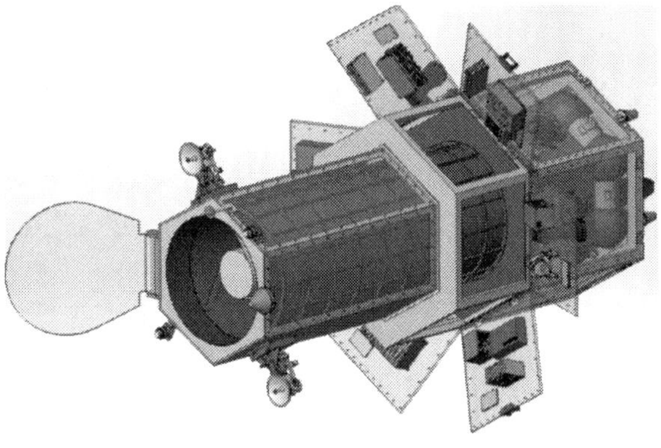

Figure 3. LRSS assembly model

When all models of on-board systems, spacecraft construction and assembly model are formed, the spacecraft weight report is developed. At this stage, balancing and inertia characteristics are determined. It is sensible to use here CAD systems, as they include tools for automated determination of those characteristics based on given values of structural elements' density.

3.5 Choice of launcher and spacehead model development

First, for the choice of launcher a design engineer has to take into account its payload capacity and dimensions of payload zone. Then, statistical data on various launchers are compared with regard for vehicle launch cost, and choice is made by criterion of minimum cost.

At this stage of the project, the spacecraft designed is placed under payload fairing (PLF) and animation picture of PLF jettison and spacecraft transfer to operational mode is realized. At the end of design process, protocols with LRSS design structure and characteristics are formed.

4 Conclusion

As it is stated [1, 2], project designs are core items of engineering degree curricula. As far as the programme "Manned and Unmanned Spacecraft and Space Systems" is concerned, the core project is "Calculation of main parameters and generation of land remote sensing satellites conceptual design based on regard to required efficiency indices".

That activity is essential for students. The reason is that it prepares them for their graduation thesis and, generally, for their future profession. As an additional incentive, our Department holds a best project contest.

Effectiveness of the engineering degree programme "Manned and Unmanned Spacecraft and Space Systems" is confirmed since 2008 by certificates of AEER and ENAEE.

References

[1] Criteria and procedure for professional accreditation of engineering programmes (AEER), URL `http://www.aeer.ru/`

[2] EUR-ACE® Framework Standards and Guidelines, ENAEE, URL `https://www.enaee.eu/eur-ace-system/standards-and-guidelines/#standards-and-guidelines-for-accreditation-of-engineering-programmes`

[3] A. Kucherov, V. Kurenkov, *Use of cluster analysis for development of star tracker mass statistical model*, Procedia Eng., **185**, 227 (2017)

[4] V. Kurenkov, *Basic principles of electrooptical survey satellites design* (Samara University Publishing House, Samara, 2020)

[5] A. Kirilin, R. Akhmetov, V. Kurenkov et al., *Generation of land remote sensing satellites conceptual design based on regard to required efficiency indices*, Procedia Eng., **104**, 65 (2015)

[6] M. Moldabekov, D. Akhmedov, S. Yelubaev et al., *Development of star tracker for satellite*, in Proceedings of the 3rd International Conference Scientific and Technological Experiments on Automatic Space Vehicles and Small Satellites, 9–11 September 2014, Samara, Russia, 107 (2014)

[7] S. Garbuk, V. Gershenzon, *Space systems for land remote sensing* (A and B Publishing House, Moscow, 1997)

[8] R. Schowengerdt, *Remote sensing: models and methods for image processing* (Elsevier, Amsterdam; Academic Press, Burlington, 2007)

[9] A. Chandra, S. Gosh, *Remote sensing and geographic information systems* (Alpha Science, 2006)

[10] V. Lebedev, *Space problems of Earth remote sensing*, Herald of the Russian Academy of Sciences, **78**, 136 (2008)

[11] Ye. Somov, S. Butyrin, A. Butko, *Satellite gyromoment guidance at area land survey based on sequence of the scanning observation courses*, in Proceedings of the 18th Saint Petersburg International Conference on Integrated Navigation Systems, 30 May—1 June 2011, Saint Petersburg, Russia, 323 (2011)

[12] V. Kurenkov, A. Kucherov, *Methodology and Software for Estimating Target Efficiency of Land Remote Sensing Satellites at the Stage of Design*, J. Phys.: Conf. Ser., **1368**, 042037 (2019)

On the main approaches to creating conditions for training personnel with higher education on an interdisciplinary basis

Nikolai Kushchev[1,*]

[1]Department "Personnel Management", Moscow Aviation Institute (National Research University), 4, Volokolamskoe shosse, Moscow, 125993, Russia

Abstract. The paper reviews the regulatory issues in the field of higher education as related to the improvement of personnel training quality based on enhancing the classification of professional educational programs in the Russian Federation. The features of the current state of higher education in Russia, the problems and possible prospects for transformations in this area, as well as the main stages of fulfilling the task of transforming the higher education professional educational programs are discussed. The liberalization of federal state educational standards of higher education and the increased independence of higher education institutions during the development of educational programs taking into account professional standards indicate the need to create an updated classification system for the Russian education. The classification system being developed should become a common one for all types of professional educational programs, including professional training programs and additional professional programs. The paper proposes new approaches to the system structure and basic classification levels of the List of specialties and areas of training, as to the establishment of the conditions for training personnel with higher education on an interdisciplinary basis, taking into account the common subject content of professional activities and providing the conditions for the existence of enlarged groups of specialties and areas of training.

1 Introduction

The issues of improving the quality of training of specialists with higher education in the Russian Federation are periodically reviewed in the context of updating the lists of higher education specialties and areas of training (hereinafter referred to as the Lists of Specialties). The lists belong to the primary documents that determine the structure of staff training for the labor market and directly influence the improvement of main professional educational programs. The task of updating these lists is caused by their incomplete compliance with the current qualifications system in Russia, one of the most important elements of which is the National Qualifications Framework of the Russian Federation [1].

Currently, the registry of the Russian Ministry of Labor and Social Protection [2] includes more than 2,000 professional standards, reflecting the view of professional communities and employers concerning the existing occupational and qualification structure of employment in the labor market. However, for certain specialties and areas of training, there are no professional standards present in the registry, and, what is more, the development is not even

*e-mail: kyshev@mai.ru

planned for some of them. This can be treated as a sign of the fact that the corresponding specialties/training areas are outdated and need to be excluded from the mentioned lists.

At the same time, there are professional standards relevant to the industries, which have no corresponding specialties and areas of training in the lists. Thus, they should be included in the lists and their development should be arranged accordingly. Therefore, it is reasonable that the President of the Russian Federation Vladimir V. Putin dedicated part of his annual Address to the Federal Assembly of April 21, 2021, to the social policy and economics issues. The head of the state noted that there are all the necessary conditions for the implementation of goals aimed at breakthrough development within the education system: "… For this, our country has a reliable potential created during the previous years in … science, education, industry …" [3]. The analysis of the social and economic development of the country confirms the conclusions made by the President. This became possible partially thanks to the implementation of national projects as a set of mechanisms and tools to improve the quality of public administration, which enabled to ensure the achievement of socially significant results in the field of social policy and education.

An evident example of this is the "Education" National Project, designed for 6 years, which was adopted on December 24, 2018 [4]. Currently, this project includes 10 federal projects. Two main goals have been approved in the project data sheet and stated as follows: the first one is to ensure the global competitiveness of the Russian education and the entry of the Russian Federation into the top 10 countries in the world in terms of the quality of general education; the second one is the upbringing of harmoniously developed and socially responsible individuals on the basis of the spiritual and moral values of the peoples of Russia, their historical, national, and cultural traditions. In accordance with the urgent need to implement the strategic development goals set by the head of our state and documented within the framework of the ongoing "Education" National Project, the main direction has been determined in terms of achieving a new high-quality level of engineering personnel training, which is associated with the necessity for the Russian education to enter the world market and ensure significant economic growth of the country.

In this regard, the task of improving the quality of training for specialists with a higher education in the field of engineering, which primarily implies aligning the internal personnel training rules with the international requirements, has become a priority not only for public administration bodies, but also for the entire scientific and pedagogical community of the country. It was then reasonably reflected by the Instruction of the President of Russia stated as follows: "… Ensure revising the list of higher education specialties and areas of preparation, as well as the nomenclature of scientific specialties for which academic degrees are awarded. Take measures to enlarge them, create conditions for training personnel with higher education and conducting scientific research on an interdisciplinary basis.

Take into account the need to preserve the peculiarities of training personnel according to specialist's programs for certain sectors of the economy" [5]. Therefore, the key task in transforming and updating the current Lists of Specialties is to prepare sound conceptual and methodological approaches and establish correlations between qualifications in the higher education system (in accordance with the Federal State Educational Standards of Higher Education) and qualifications that are obvious for the labor market (in accordance with professional standards).

2 Analysis of the state of educational programs classification and the prospects for the transformation of the existing lists of higher education specialties and areas of training

Before proceeding to the substantiation of the need for transformation of the existing higher education Lists of Specialties, it is advisable to review the conditions, under which the current classification system was created in Russia, and its current state. The analysis shows that the current higher education Lists of Specialties were introduced by orders of the Russian Federation Ministry of Science and Higher Education No. 1061 dated September 12, 2013 "On the approval of the lists of specialties and areas of training of higher education" (as amended and supplemented) and No. 1060 (as of September 28, 2020) "On the approval of the lists of specialties and areas of training of higher education applied in the implementation of higher education programs, containing information constituting a state secret or official information of limited distribution". We would like to draw attention to the fact that the composition of the list items in this edition was identical to that of the 2009 edition. During the lists development, a different classification system, consisting of 57 enlarged groups (instead of the 29 previously existing groups), including a certain set of specialties and areas of training (hereinafter referred to as the Enlarged Groups) was taken as the basis. This model was structurally close to the then-current International Standard Classification of Education (ISCED-2011 version), which provided for 25 combined groups, containing educational fields. The version currently applied is ISCED 2013, providing for 29 directions, which include the available specialties.

The distribution of specialties (areas of training) into groups historically developed back in the years of the Soviet Union. This is how all lists and classifiers were created, including those for higher professional education. It is worth noting that such large groups were formed exclusively based on their belonging to an industry. Such economy-relevant statistics as the admission of applicants and the graduation of specialists from educational institutions, which, by the way, also mainly belonged to an industry sector, were formed accordingly by sectoral groups. This made it possible to compare them with population employment indicators and, consequently, to assess the staffing needs of each of the industries, etc. More detailed information was provided per professional groups (into which the enlarged groups were subdivided). These data were used to make decisions on the adjustment of student admission and training state plans, as well as on the implementation of the educational process, the organization of training for new professions and study areas, etc.

The current coding structure does not meet the modern requirements and does not comply with the international practice. Nowadays, a more effective system for coding educational programs is considered to be the one that includes two parallel directions: the first direction is the classification of educational programs per fields of professional activities, and the second one is per levels and types of education. So, while the ISCED-2011 version applied a three-level hierarchical subsystem of horizontal classification, including "Extensive fields", "Narrow fields", "Detailed fields", the ISCED-2013 version already uses the "educational and professional training fields" concept, within which the "scientific area", "direction", and "specialty" levels are defined [6]. In addition, as a well-known expert in the field of classification system of educational programs Sazonov B.A. mentions, the proposed coding system for the existing lists "… contains only one level of classification—"enlarged groups", which has proven itself inapplicable for the presentation of statistical data: who would be interested to know the total number of specialists being trained for "geodesy and oil and gas business activities" without the possibility to provide the specific numbers for the constituent professional areas.

Only two capabilities remained for the educational statistics: to present the aggregate statistical data for the educational system levels in general or to provide them separately for each of more than 1,200 directions of training, specialties and professions, and this is exactly what we see in the modern Russian reference books containing educational statistics" [7]. This largely explains the reason why instead of a single list of higher education specialties and areas of training in the Russian Federation, there are currently 11 uncoordinated lists developed for individual levels of higher education, which actually doubles the total number of areas of training and specialties. Such an increase in the number and volume of the lists is related to the creation of a simplified coding system for educational programs. The simplification idea was not to create a separate description of the coding system, but to assign a code to each area of training possible for implementation immediately in the list. As a result, the same training area name, for example, "Mechanical Engineering", appears not in one, but in three lists (for bachelor's, master's, and postgraduate programs), and each time with a new code (15.03.01, 15.04.01, 15.06.01). Thus, an attempt to create a simplified coding system has led to the "multiplication" of lists and the total number of areas of training in them. It should be noted that the current Lists combine the traditions of national education and are already harmonized with the following systems: the International Standard Classification of Education (ISCED-2013) adopted by UNESCO and the International Classification of Science and Technology (FOS-2007) effective in the OECD (Organization for Economic Cooperation and Development) countries.

3 The proposed structure and basic classification levels of the new List of specialties and areas of training of higher education

In our opinion, the current legislation in the field of education in Russia, in general, provides the necessary quality of professional personnel training; still the existing set of regulatory documents may well become the basis for the transformation and revision of the Lists of specialties and areas of training of higher education. While working on the implementation of the presidential instructions related to the development of the concept for consolidating the Lists of Specialties, the Ministry of Education and Science of Russia, the Coordination Councils for the fields of education, Federal educational and methodological associations, as well as the entire pedagogical community and training customers (employers) will have to solve a number of additional tasks: to analyze and clarify the correspondence tables between the lists of federal state educational standards of higher education (hereinafter referred to as Federal Standards) and professional standards, as well as to align the corresponding lists with the professional standards registry of the Ministry of Labor and Social Protection of Russia; to make decisions on the introduction of new areas of training and specialties, taking into account the increased technological level of modern production; to exclude the items not demanded by the labor market and the educational service customers from the Lists of Specialties; to develop a unified system of classification and coding of main professional education programs for all levels of education, which would include additional professional training, retraining, and advanced training programs; to assign codes to transformed lists of training areas and specialties, using the developed classification system; and to obtain approval for the updated lists.

In addition, the analysis of the nomenclature of qualifications included in the Ministry of Labor and Social Protection registry reveals their incompliance with the qualifications, which are assigned to the graduates of educational institutions in accordance with the approved Lists of specialties and areas of training of higher education, making them opaque for the labor market and raising a question of what labor functions the university graduates are trained to perform [8]. At the same time, the leading organizations, which are the customers of training in our country (employers), are beginning to understand more precisely what kind of specialists they need, and are increasingly expressing their views on this issue. In this regard, the existing conditions have indeed predetermined the emerging need for the transformation of the current higher education Lists of Specialties.

The fulfillment of these tasks is facilitated by the fact that the current higher education Lists of Specialties and International Classifiers contain three levels. Besides, the multiplicity of lists can be eliminated by abandoning the above-mentioned simplified coding system. The List of Specialties should not contain any repeating names of specialties and areas of training. Thus, the number of items in the list will be reduced by almost half (by 38%) and, most importantly, the continuity of the main educational programs throughout the levels of higher education will be evident. Therefore, the structure of the new list draft should retain three levels of classification, namely: field of education—Enlarged Group—area of training (specialty). It is advisable to conduct the coding of educational programs per the levels of higher education not in the List of Specialties, but in the All-Russian Classifier of Education Specialties. It should be noted that the International Classifiers ISCED and FOS are uniform for all levels of higher education, which enabled to make them concise and define the continuity of educational programs throughout the corresponding levels of higher education. For clarity, let us compare the main indicators of the current lists and the proposed innovations:

1. The current state in the field of classifying higher education specialties and areas of training: there are 11 uncoordinated lists; within the lists, a code is assigned to each area of training/specialty possible for implementation (for higher education, there are 456 area/specialties in 56 Enlarged Groups); the lists correspond to the traditions of national education and are consistent with the International Standard Classification of Education (ISCED-2013) and the International Classification of Science and Technology (FOS-2007); the current lists of areas of training and international classifications are composed of three levels.

2. A single list of higher education specialties and areas of training, which will ensure the following: preservation of the three-level classification; a new structure with the "field of education—Enlarged Group—area of training (specialty)" format; the classifier is the same for all higher education levels, making it more concise (for higher education, there will be 284 items in 56 Enlarged Groups, meaning a total reduction by 38%); the coding of educational programs per levels of higher education will be carried out in the All-Russian Classifier of Education Specialties; maintaining consistency with the International Standard Classification of Education (ISCED-2013) and the International Classification of Science and Technology (FOS-2007).

At the same time, it is necessary to take into account that various proposals for the consideration of other classifiers and lists that can be applied as a basis may arise in the process of expert discussion. For instance, these could be the International Classifier ISCED-2013 or the Order of the Ministry of Labor of Russia No. 667n dated September 29, 2014, which undoubtedly need to be considered when developing the new List of Specialties. Still, none of them can be taken as the only basis, since they were created to solve other specific tasks. Moreover, the ISCED-2013 classifier is intended only for the collection of statistical data for UNESCO and is used in countries with very different levels of economic and educational development. However, at the same time, it is not focused on its application for the education systems and economies of the leading countries around the world. Therefore, it is very important that while forming the new List of Specialties, the Ministry of Education and Science of the Russian Federation should adopt a "soft" version of its alignment with ISCED-2013, taking into account the national needs of training personnel for our economy. As an example, it could be mentioned that "Information and communication technologies" are named as a separate area of education (science), which contains only three areas of training as follows: "Computer usage", "Database creation", and "Software development". It is obvious that our system of training personnel in the field of computer science, information technology, information security, and management of engineering systems possesses generally recognized global leadership, and it must be preserved and developed to meet new challenges. Computer science cannot be separated from mathematics forming its foundations, and information technology and management of engineering systems should not be removed from the engineering field of training. Moreover, the recommendations for the application of ISCED 2013 explicitly state that it can be applied at the national level "in those countries that have not yet developed their own standard classification of educational fields", meaning the third world countries. Moreover, they further state that "those wishing to apply this classification within their country will need to adjust certain groups of areas for them to match better the scope and/or range of training areas in accordance with their national needs" [6].

4 Creation of the conditions for training personnel with higher education on an interdisciplinary basis

It should be noted that if a single List for the entire higher education is formed and the system of coding and accounting of educational programs is changed, then this will become the basis for the further development of interdisciplinary educational programs, including within the framework of implementing educational programs in a network format. It is related to the fact that even now the Federal State Educational Standard of Higher Education 3++ and the current lists of areas of training (specialties) allow universities to implement any interdisciplinary educational programs within the framework of independently established training profiles and specializations. At the same time, it is advisable to assign the functions of collecting and systematizing information on the profiles (specializations) to the Federal Education and Methodology Associations or to the Coordination Councils for the fields of education. At the same time, this really supports the implementation of the "2+ system" as follows: the possibility of organizing a unified training for students within a common Enlarged Group at a university during the first two years of studies and providing the students with the right to choose the area of training (specialty) upon completing their second year; a single list of admission examinations (USE); uniform requirements for the training performance conditions and professional "core" (uniform general professional competencies and their achievement indicators); a flexible mechanism for choosing educational paths.

At the same time, in order to implement the requirements for preserving the peculiarities of training personnel under the specialist's programs, which is especially important in the engineering field of education, one should take into account such an important issue raised by the employers as the lack of university graduates trained particularly under the specialist's programs. Such highly demanded training is limited by the current Lists of specialties and areas of training of higher education. The number of engineering specialist's areas of training in the Lists is many times lower than the number of training areas under bachelor's programs. Currently, it is the Lists that determine the possible paths for obtaining higher education in specific engineering areas. When they were formed in 2013, there was a strict requirement that specialties and areas of training could not have a common basis or a common name. As time has shown, this requirement has led to an unreasonable limitation of the possibilities to meet the needs of the economy sectors in personnel with appropriate qualifications.

The introduction of a unified list of higher education specialties and areas of training will allow to: ensure the development of interdisciplinary educational programs, including within the framework of implementing educational programs in a network format; introduce training under specialist's programs for an arbitrary training area within the framework of the Enlarged Group if such training is necessary; eliminate the strict requirement for the mandatory separation of specialty and area of training within a common basis and a common name; ensure the unification of personnel training during the first two years of studies within an individual Enlarged Group, including the unification of bachelor's and specialist's programs. Thus, the development of a unified classification List of specialties and areas of training of higher education, which is based on an interdisciplinary nature of training, support of the network forms of primary educational programs organization, implementation of the "2+ system"; provision of a flexible educational paths mechanism, and assurance of the possibility of introducing a specialist's program in any area of training along with the unification of the first two years of studies, makes it possible to further harmonize the contents of the fourth generation Federal State Educational Standard concept. In this regard, for the more effective implementation of training and the development of a new List, it is advisable to update the composition of the enlarged groups of specialties and areas of training, which, in turn, requires the formation of criteria defining the need for individual Enlarged Groups. This is necessary to make the process of forming the List an objective one. The existence of separate enlarged groups will allow to: ensure the possibility of organizing unified training for students within an individual Enlarged Group during their first two years of studies and providing the students with the right to choose the area of training (specialty) upon completing their second year; use of a single list of admission examinations; and develop uniform requirements for the training performance conditions and professional "core" (uniform general professional competencies and their achievement indicators).

References

[1] Order of the Ministry of Labor and Social Protection of the Russian Federation No. 148n dated April 12, 2013 "On the approval of qualification levels for the purpose of developing professional standard drafts"

[2] Order of the Ministry of Labor and Social Protection of the Russian Federation No. 667n dated September 29, 2014 "On the registry of professional standards (list of types of professional activities)"

[3] Address of the President of the Russian Federation to the Federal Assembly dated April 21, 2021 "Address of the President to the Federal Assembly" [Electronic source], URL http://www.consultant.ru/document/cons_doc_LAW_382666/ (access date: 2021-05-07)

[4] *Nine assignments of Minister O. Vasilyeva*, Rossiyskaya Gazeta, URL https://rg.ru/2018/07/31/olga-vasileva-raskryla-podrobnosti-nacproekta-obrazovanie.html (access date: 2021-04-21)

[5] Instruction of the President of the Russian Federation dated March 28, 2020 No. Pr-589 (p.1 zh-2)

[6] International Standard Classification of Education 2013, ISCED 2013 (Institute of Statistics, UNESCO, 2014)

[7] B.A. Sazonov, *Classification of professional educational programs: current state and possible prospects*, Higher Education in Russia, **11(217)**, 20 (2017)

[8] B.A. Sazonov, V.E. Yatsenko, Yu.V. Girinovich, *All-Russian classifiers in the field of education: creation, maintenance, development* (FIRO, Moscow, 2006)

Features of training of marine ship radio specialists

Elena Lavrenteva[1,*]

[1]Federal State Educational Institution of Higher Education "Admiral Makarov State University of Maritime and Inland Shipping", 5/7, Dvinskaya st., Saint-Petersburg, 198035, Russia

Abstract. The article discusses the features of training radio specialists for sea transport, taking into account international and national requirements for the qualifications of seafarers. It is emphasized that only persons who have diplomas and qualification certificates of the established form are allowed to occupy the positions of ship's crew members. As part of the study, qualification documents were systematized for a GMDSS limited area operator, GMDSS operator, GMDSS second class radio electronics, radiotelephone operator and GMDSS first class radio electronics. The areas, types and tasks of the professional activity of the graduates of the educational program "Radio communication and electrical radio navigation of the marine fleet" have been determined. Among the main objects of professional activity, radio-electronic equipment on ships and coastal centers of sea and water transport is highlighted. It was emphasized that the results of training specialists are assessed by the acquired universal, general professional and professional competencies, i.e. the ability to apply knowledge, skills and personal qualities in accordance with the tasks of professional activity. The article examines modern trends in determining the quality of education by using internal and external assessments. It is concluded that the training of radio specialists for maritime transport in modern conditions is of great importance, especially taking into account the development of transportation along the Northern Sea Route and the industrial development of the Arctic. At the same time, the defining condition is the inclusion of international and national requirements in the educational process, then graduates in the field of radio communication and electric radio navigation of the marine fleet will be in demand in the Russian and foreign freight market.

Maritime transport issues are of great importance both in scientific literature [1–6] and in international and national regulatory documents. Only persons holding diplomas and qualification certificates of established standard are allowed to hold positions of ship crew members. Such procedures are defined by the International Convention on Standards of Training, Certification and Watchkeeping for Seafarers, 1978 [7] and the Order of the Russian Ministry of Transport [8]. Thus, ship radio specialists, like all members of ship's crews, in addition to the relevant professional education, must undergo certification, receive and confirm qualification documents with the established frequency:

- to the operation level: operator of the limited region of GMSSB, operator of GMSSB, radio electronics of the second class GMSSB, operator-radio telephone operator;
- to the management level—radio electronics of the first class GMDSS.

*e-mail: e_lavrentieva@mail.ru

The Global Maritime Distress and Safety System (GMDSS) was developed by the International Maritime Organization (IMO) and is an international system using modern terrestrial, satellite and shipborne radio communication systems.

The documents required for certification of ship radio operators are systematized in table 1.

Table 1. Documents for certification of ship radio specialists

No.	Position/Documents
1	Common documents for all positions: • identity document; a document confirming fitness for work on ships for health reasons; • certificates of: initial safety training; training in lifeboats and rescue boats and rafts; preparation for fire fighting; • training in accordance with Regulation of the STCW Convention; security training
2	Operator-radiotelephone operator: • certificate of completion of training for radiotelephone operators under a program that meets the requirements of the STCW Convention
3	GMDSS restricted area operator: • a document on obtaining higher or secondary vocational education in the field of navigation; • certificate of completion of GMDSS restricted area operator training
4	GMDSS operator: • a document on obtaining higher or secondary vocational education in the field of navigation; • certificates of sailing on ships for at least six months with the fulfillment of radiocommunication duties; • certificate issued by a maritime educational organization on the completion of GMDSS operator training in accordance of the STCW Convention
5	Radioelectronics of the second class GMDSS: • a document confirming the receipt of higher or secondary vocational education in the field of radio communication or in the field of navigation; • documents confirming the implementation of training programs, or certificates of sailing, confirming the experience of sailing at least six months with the performance of duties in radio communications; • certificate of completion of GMDSS second class radio electronics training
6	Radioelectronics first class GMDSS: • a document on obtaining higher professional education in the field of radio communication or in the field of navigation; • diploma in radio electronics of the second class of GMDSS; • certificate of sailing for at least 24 months in the position of second class radio electronics

Graduates prepare for professional careers in the following areas:

- communications, information and communication technologies (in the areas of: provision and maintenance of transport activities; technical operation of transport facilities; information and telecommunication technologies in transport);
- transport (in the areas of: provision and maintenance of transport activities; technical operation of transport facilities; information and telecommunication technologies in transport).

Table 2. Types of professional tasks of graduates

No.	Types and tasks of professional activity
1	Operational and technological type of tasks Tasks: • ensuring the uninterrupted operation of electronic systems; • monitoring and determining the operability of installed, operated and repaired transport radio-electronic equipment, predicting its technical condition; • technical operation of transport radio-electronic equipment in accordance with the requirements of normative and technical documentation; • support of work on the modernization of transport radio-electronic equipment, subject to agreement with the developer of the equipment, selection and replacement of its elements and systems; • marketing of services for the operation of transport radio-electronic equipment of various forms of ownership; • selection of security alarm systems, implementation, commissioning of security systems; • technical operation of communication systems and electrical radio navigation in maritime transport in accordance with the requirements of the STCW Code
2	Organizational and managerial type of tasks Tasks: • ensuring the uninterrupted operation of radio electronic systems; • participation in the implementation of the functions of supervision over the safe operation of transport radio-electronic equipment; • conducting an internal audit and preparing certification of objects of technical operation of transport radio-electronic equipment

The educational program is focused on solving operational, technological and organizational and managerial types of tasks presented in table 2.

Among the main objects (or areas of knowledge) of the professional activity of graduates, radio-electronic equipment on ships and coastal centers of sea and water transport stands out: radar, radio navigation and communication systems, systems and means for monitoring and diagnosing the technical condition of the radio-electronic equipment in operation, systems for transmitting information about the movement of vehicles and external conditions of their operation, systems for integrated processing, display and registration of information about the movement of vehicles and external conditions, systems for controlling the movement of ships and warning systems for their dangerous approaches, radio communication systems and electric radio navigation of maritime transport.

The results of mastering the educational program are determined by the competencies acquired by graduates, i.e. the ability to apply knowledge, skills and personal qualities in accordance with the tasks of professional activity. Highlighted are universal, general professional and professional competencies.

Universal competences are the same for all educational standards for bachelor's and specialist's degrees:

- the ability to carry out a critical analysis of problem situations on the basis of a systematic approach, to develop an action strategy;
- ability to manage a project at all stages of its life cycle;
- the ability to organize and manage the work of a team, developing a team strategy to achieve a set goal;

- the ability to use modern communication technologies, including in foreign language(s), for academic and professional interaction;
- the ability to analyze and take into account the diversity of cultures in the process of intercultural interaction;
- the ability to determine and implement the priorities of their own activities and ways to improve them on the basis of self-esteem and education throughout life;
- the ability to maintain the proper level of physical fitness to ensure full-fledged social and professional activity;
- the ability to create and maintain safe living conditions, including in the event of emergencies and military conflicts;
- ability to use basic defectological knowledge in social and professional spheres;
- the ability to make informed economic decisions in various areas of life;
- the ability to form an intolerant attitude towards corrupt behavior.

General professional competences (GPK) determine training in the engineering field, which is shown in table 3.

Table 3. General professional competencies

GPC Group	General professional competencies (GPC)
Natural and scientific laws of the modern world and general engineering field	Ability to use the basic laws of mathematics, units of measurement, fundamental principles and theoretical foundations of physics, theoretical mechanics
Legal literacy	Ability to apply the foundations of Russian and international legislation in the field of professional activity
Information and communication technologies	Ability to present information in the required format using information, computer and network technologies, taking into account the basic requirements of information security
Engineering graphics	Ability to carry out measurements and instrumental control, process results and evaluate errors
Environmental safety and labor protection	Ability to apply technical means and technologies to minimize negative environmental consequences, ensure safety and improve labor in the field of professional activity
Modeling complex systems	Ability to apply the fundamental foundations of modeling theory as the main research method and a scientifically grounded method for assessing the characteristics of complex systems used for decision-making in various areas of professional activity

Professional competencies reflect the specifics of training radio specialists for marine vessels. The object or area of knowledge is radio electronic equipment on ships and coastal centers of sea and water transport. The relationship between the tasks of professional activity and professional competencies of graduates is systematized in table 4.

Modern trends in determining the quality of graduate training, incl. and marine ship radio specialists are developed using internal and external evaluation. As part of an internal assessment, students consider the conditions, content, organization of the educational process as a whole and individual disciplines and practices. Reviewing of educational programs is

Table 4. Professional competence of marine ship radio specialists

The task of professional activity	Professional competence
Operational and technological type of professional activity	
Security uninterrupted operation of electronic systems	Ability to commission, maintain and repair radio electronic systems
Monitoring and determining the operability of the installed, operated and repaired transport radio-electronic equipment, forecasting its technical condition	Able to test and determine operability of the installed, operated and repaired transport radio electronic equipment, predict its technical condition
Technical operation transport radio-electronic equipment in accordance with the requirements of regulatory and technical documentation	Ability to exercise Technical operation of radio communication equipment and electrical radio navigation in accordance with the requirements of regulatory and technical documentation
Maintenance of work on the modernization of transport radio-electronic equipment, subject to agreement with the developer of the equipment, selection and replacement of its elements and systems	Ability to participate in the modernization of marine radio-electronic equipment, form recommendations for the selection and replacement of its elements and systems
Carrying out marketing of services for the operation of transport radio electronic equipment of various forms of ownership	Ability to express competent judgments on the basis of marketing services for the operation of transport radio-electronic equipment of various forms of ownership
Selection of security alarm systems, commissioning and commissioning of security systems	Ability to make an informed choice of security alarm systems, perform commissioning of security systems
Technical operation of communication and electrical radio navigation systems in maritime transport in accordance with the requirements of the STCW Code	Ability to carry out the technical operation of communication systems and electrical radio navigation in maritime transport in accordance with the requirements of the STCW Code
Organizational and managerial type of tasks of professional activity	
Ensuring the uninterrupted operation of electronic systems	Ability to carry out organizational and methodological support for the technical operation of radio electronic systems
Participation in the implementation of the functions of supervision over the safe operation of transport radioelectronic equipment	The ability to organize safe conditions for the installation and commissioning of transport radioelectronic equipment
Internal audit and preparation of certification of objects of technical operation of transport radio electronic equipment	Ability to solve problems of internal audit and preparation of certification of objects of technical operation of transport radio-electronic equipment

carried out; development of objective procedures for assessing the level of knowledge and skills of students, competencies of graduates; ensuring the competence of the teaching staff; regular self-examination with the involvement of representatives of employers; informing the public about the results of their activities, plans, innovations. The current control of the progress, intermediate and final certification of students is carried out by using the funds of evaluation means, which allow assessing the achievement of the learning outcomes planned in the educational program and the level of formation of all declared competencies.

In order to bring the current monitoring of progress and intermediate certification of students closer to the tasks of their future professional activities, managers and employees of specialized organizations with at least 3 years of experience in this professional field, as well as teachers of related educational fields, are involved in these procedures.

- An external assessment of the quality of educational activities is carried out to confirm it to educational standards, and can be carried out within the framework of professional and public accreditation conducted by employers, their associations, as well as organizations authorized by them, including foreign ones. It is possible to implement it by authorized national professional and public organizations that are part of international structures in order to recognize the quality and level of training of graduates that meet the requirements of professional standards and the labor market.
- The training of radio specialists for maritime transport in modern conditions is of great importance, especially taking into account the development of transportation along the Northern Sea Route and the industrial development of the Arctic. At the same time, the defining condition is the fulfillment of international and national requirements in the educational process, then our graduates will be in demand in the Russian and foreign freight market.

References

[1] A. Brodje, M. Lundh,, J. Jenvald, J. Dahlman, *Exploring non-technical miscommunication in vessel traffic service operation*, Cogn. Technol. Work, **15**, 347 (2013), DOI: 10.1007/s10111-012-0236-5

[2] A.N. Liashuk, S.A. Zavyalov, E.A. Chashin, *Small monopole transceiver antenna for medium frequencies*, in 17th International Conference of Young Specialists on Micro/Nanotechnologies and Electron Devices (EDM), 30 June—4 July 2016, Erlagol, Russia (2016), DOI: 10.1109/EDM.2016.7538712

[3] I.V. Dulkeyt, S.A. Zavyalov, E.A. Chaschin, A.R. Shigabutdinov, *Some aspects of providing information security of hydrocarbons extraction and transportation in the Arctic Zone of the Russian Federation*, Procedia Engineering, **113**, 344 (2015), DOI: 10.1016/j.proeng.2015.07.277

[4] I.F. Bazhukov et al., *Prospects of using medium-wave band for radio communication with rescue mobile teams of EMERCOM*, J. Phys.: Conf. Ser., **944**, 012010 (2018)

[5] R. Cwilewicz, J. Lisowski, *The integrated maritime education and research activity of Gdynia Maritime University for seafarers*, in The 12th Annual General Assembly of IAMU—Green Ships, Eco Shipping, Clean Seas, 12–14 June 2011, Gdynia Maritime University, Gdynia, Poland (2011)

[6] J. Mansson, M. Lutzhoft, B. Brooks, *Joint Activity in the Maritime Traffic System: Perceptions of Ship Masters, Maritime Pilots, Tug Masters, and Vessel Traffic Service Operators*, J. Navigation, **70(3)**, 547 (2017), DOI: 10.1017/S0373463316000758

[7] International Convention on Standards of Training, Certification and Watchkeeping for Seafarers (1978)

[8] Order of the Ministry of Transport of Russia dated 15.03.2012 No. 62 (rev. dated 13.05.2015), "On approval of Regulations on Certification of Crew Members of Sea Vessels"

Talent cultivation method of aerospace manufacturing engineering incorporating new aerospace technology

Rongqiang Liu[1,*], *Guanxin* Chi[1], *Fei* Wang[1], *Lijun* Yang[1], *Honghao* Yue[1], and *Yifan* Lu[1]

[1]Department of Aeronautics and Astronautics Manufacturing Engineering, School of Mechatronics Engineering, Harbin Institute of Technology, Harbin, 150001, P.R. China

Abstract. In order to meet the needs of the state and society for improving the quality of undergraduate education and cultivating innovative talents, individualized training mode has gradually become the direction of higher education reform. In China, there is a long-standing situation that talent cultivation is out of touch with industrial demand. In order to explore the training mode of innovative talents in the new era, the idea of cultivating individualized talents with scientific research as feedback to teaching is established in this paper. The latest research results of aerospace are incorporated into the training process of aircraft manufacturing engineering professionals and professional knowledge system are optimized. By building high-level courses with overseas scholars and domestic industry experts, an off-campus practice training base for personalized talent training is established, and the teaching contents and methods is reformed. The practice has proved that the personalized talent training mode and method proposed in this paper have achieved good results.

1 Development and evolution of the major

The foundation of aircraft manufacturing engineering major in Harbin Institute of Technology (HIT) is the automated mechanical design and manufacturing major of HIT, and its origin can be traced back to 1958. Harbin Institute of Aviation Technology was merged into HIT to set up the department of aviation engineering. In 1963, the department of aviation moved out to support the professional construction of several national characteristic colleges and universities, then this major was closed. In 1993, it began to prepare for the establishment of new postgraduate training in the aerospace manufacturing engineering discipline, and it has been 17 years since 2004. To promote the construction and development of aircraft manufacturing engineering major, the Department of Aerospace Manufacturing Engineering was established in 2008. In 2009, this major became a key discipline in Heilongjiang Province, and in 2019, it entered the provincial first-class professional construction list. In 2021, it entered the national first-class professional construction list. The major, 'based on aerospace and serving national defence', is a multidisciplinary integration major integrating aircraft manufacturing technology, computer application technology, and mechatronics technology. Meanwhile, this major trains compound senior professionals for the design and manufacture of mechanical and electrical products, production organization management, basic theory, and applied technology research and development in aerospace and related fields.

*e-mail: liurq@hit.edu.cn

2 Teaching staff and research direction

There are 42 full-time backbone teachers in this major, including 19 professors, 8 associate professors, and 9 lecturers. Among them, there is one academician of the Chinese Academy of Engineering, one national famous teaching teacher award of colleges and universities, 11 people who were selected into the national talent support program, and 18 people who have studied abroad for more than one year. There are several teaching and research platforms, such as the National Defence Key Discipline Laboratory of Aerospace Space Agency and Control Technology, the National Innovation and Intelligence Introduction Base of Equipment Design and Manufacturing Science and Technology, and the Joint Laboratory of Space Machinery Science and Technology of China Academy of Space Technology—Harbin Institute of Technology.

This major has many strong professional directions, such as planetary surface movement systems, large-scale space folding, and exhibition mechanism, space docking and unlocking, planetary surface sampling and detection, and intelligent manufacturing of equipment. In 1999, under the leadership of Professor Zongquan Deng, the professional founder, we took the lead in researching the key technologies of the lunar rover's mobile system in China, and innovatively designed eight configurations of the lunar rover's mobile suspension. Among them, the 6-wheel rocker arm type mobile configuration is determined as the mobile system configuration of China's Chang'e III "Yutu" lunar rover [1]. The invention of the horizontal pendulum linkage lunar rover transfer mechanism has realized the reliable transfer of the 2.4 m drop of the "Yutu" lunar rover, which has become a highlight of the Chang'e III mission. The configuration of an active-passive hybrid rover mobile system with a peristalsis function was proposed. It has the functions of sinking and climbing, and was adopted by the "Tianwen-I" rover. Completed the overall scheme design and key technology research of lunar surface sampling in China, and jointly developed the "Chang'e-V" drilling and sampling model product with the engineering department [2].

Figure 1. Yutu lunar rover

Figure 2. Lunar surface sampling device of Chang'e-V

We closely integrate the aerospace scientific research results with student training, so that the students can apply what they have learnt in their career and dedicate to China's aerospace industry. Meanwhile, a large number of "Yutu" lunar rover mobile system, "Zhurong" Mars rover mobile system, Chang'e V lunar surface drilling sampling and packaging system, large space extension arm, hexapod mobile robot, and large-scale low gravity simulation system are introduced in the teaching and graduation dissertation.

3 Professional and personalized talent training concept

This major is oriented to aerospace manufacturing, with distinctive national defence characteristics. In the major curriculum, the idea of "thick foundation and wide calibre" is fully implemented, the students' generous basic knowledge of natural science, technical science, and mechanical design and manufacturing is emphasized, the cultivation of generous basic theories such as mathematics, mechanics, basic theories of mechanical design and manufacturing, aircraft manufacturing technology and the ability to apply advanced and reasonable knowledge is reinforced.

The major of aircraft manufacturing engineering pays attention to the cultivation of excellent ideological quality, scientific quality, humanistic quality, and good ability to analyse, express, and solve engineering problems, and focus on the cultivation of self-study ability, innovative consciousness, practical ability, organization and coordination ability, patriotism and dedication, honesty and pragmatism, and physical and mental health.

Broaden the innovative practice teaching links and build innovative practice base clusters with aerospace engineering departments. Turn scientific research achievements into teaching content, and turn scientific research strength and discipline advantages into teaching advantages. In combination with the development of aerospace technology, the course content will be updated in real-time, and relying on specialized course laboratories in the school, the independent, appointment-based and open experimental teaching will be realized. Improve the decentralized and personalized graduation practice base construction, build the practice management and evaluation system, and explore the establishment of a school-enterprise integration practice teaching system suitable for first-class undergraduate majors. Based on professional advantageous scientific research direction, we build 1–2 scientific research and practice platforms with display degrees and fully exercise students' practical ability. Actively update laboratory equipment, strengthen the challenge and innovation of the course experiment, establish 5 national and provincial teaching practice bases.

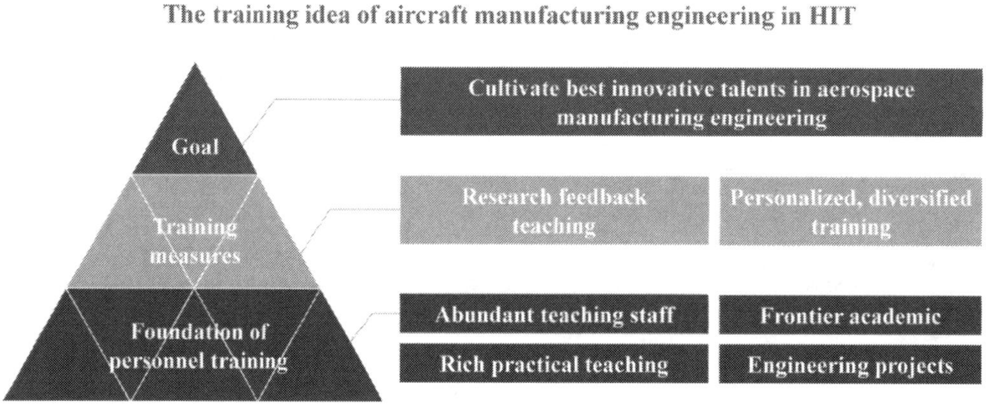

Figure 3. Training ideas of aircraft manufacturing engineering professionals in Harbin Institute of Technology

Figure 3 shows the training idea of aircraft manufacturing engineering major in HIT. Relying on HIT's abundant teaching staff, rich practical teaching foundation and scientific research advantages, the aircraft manufacturing engineering major builds a personalized training system suitable for research universities through personalized, diversified training, research feedback teaching, and other measures.

Figure 4 shows the practical teaching process of aircraft manufacturing engineering major in Harbin Institute of Technology, which embodies the training concept of "strong practice and innovation" in aircraft manufacturing engineering major.

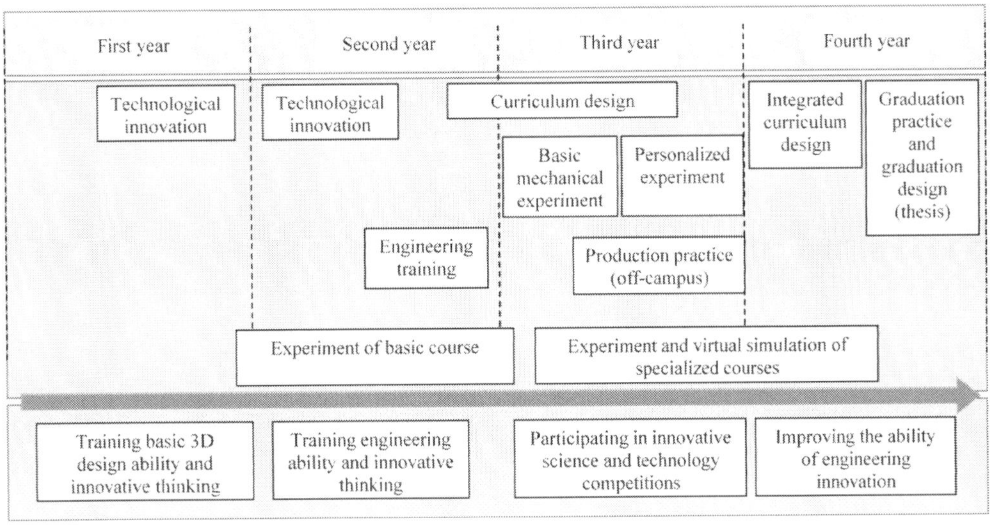

Figure 4. Practical teaching history diagram of aircraft manufacturing engineering major in Harbin Institute of Technology

Establishing the talent education concept of "being responsible for students' personalized growth and tapping students' personalized potential", and building the personalized talent training concept of research universities that "everyone has their particularity and uniqueness, and needs the expertise to show their individuality under the universal knowledge system". Establishing an undergraduate tutorial system implementation mode of mutual selection and interaction between teachers and students, from enrollment to graduation to follow-up career, with 100% students participating and significant gains.

In the aspect of curriculum system, specialized module courses are decomposed, and students choose courses according to their interests and development points; In the aspect of scientific and technological innovation practice, set up tutorial system, arrange students to enter the laboratory to follow their tutors to engage in scientific research activities in sophomore year; In the aspect of graduation practice, decentralized practice is adopted, and major aerospace units are selected as practice places. Then students can enter workshops, teams, and groups, enter the production line and participate in the actual production process.

Determining the goal orientation, adhering to the combination of talent demand orientation and traditional characteristics of running a school, and taking the related technologies of aerospace intelligent equipment system as the object, and establishing the talent training framework as follows:

1. One goal: to train high-end talents in the field of aircraft manufacturing engineering for "Made in China 2025" and future development.
2. Two ways: classroom teaching and practical training.
3. Three foundations: mechanics, electricity, and mechanics.
4. Four series: design, manufacturing, measurement and control, management course series.

4 Talent cultivation method integrating new aerospace technology

4.1 Teaching methods

On the research of professional teaching methods, the teaching link of aircraft manufacturing engineering major vigorously promotes the reform of the 'research-based learning' method. Based on professional courses with relatively independent knowledge and no lack of interdisciplinary points, several research directions under the connotation of the courses are established, with grouped research topics throughout the teaching process as the assessment object, teaching, and research are combined. Let students walk into the classroom with questions, take the realization of project research objectives as the driving force, and fully, actively acquire courses and related knowledge; Actively organize teamwork after class, use professional knowledge to carry out innovative design and practice, maximize the role of individuals in the project team, and jointly complete the project research content.

The individualized professional teaching mode of "research-based learning" requires that group practice in class be the mainline, the curriculum objectives, curriculum structure, and teaching content be optimized, and a full-cycle teaching design of "teaching preparation-teaching project design-concrete implementation-effectiveness feedback" be formed.

4.1.1 Teaching preparation

Using constructivism and metacognition theory, carrying out the investigation and analysis of students' cognitive status. Get the basic state of students' autonomous learning subjects and their subjective understanding of the planning of scientific research projects in the curriculum system, which provides the basic basis for effective teaching activities and data for teaching effect evaluation.

4.1.2 Instructional design

Setting the curriculum objectives in the 'research-based learning' system, planning the curriculum structure and organizing teaching materials in combination with teaching, materials, such as lecture notes, expanding knowledge, designing cases, typical applications and scientific research projects that students need to complete in groups, and finally, form a systematic teaching document and curriculum system.

4.1.3 Classroom teaching

Combined with the professional characteristics and academic direction of the course, group research demonstration and implementation combined with classroom teaching content are carried out. Establishing a scientific research team in the teaching cycle by self-selecting topics or grouping propositions, and establishing process management mechanisms such as internal discussion, state feedback, and teacher guidance. The examination method of the course is changed from proposition examination to project completion reply.

4.1.4 Extended development

Establishing a long-term tracking and encouraging plan for the achievements of grouped projects, selecting the research team of the teachers from the students who have completed the teaching process, and digging deep into the research direction confirmed in the course. Furthermore, students can master scientific knowledge and the steps and methods of scientific research more deeply, develop the ability of scientific argumentation and cultivate research skills.

4.1.5 Perfection of evaluation

Establishing the evaluation system of teaching effectiveness of "research-based learning", comprehensively evaluating the tightness of the combination of the teaching mode and the

Figure 5. Innovative training courses taught in groups

curriculum and its influence on students from various aspects such as students' satisfaction with classroom learning, the degree of investment in scientific research projects, the academic level of scientific research achievements, the cohesion of teamwork and self-evaluation after the completion of the curriculum.

4.2 Application of scientific research results in teaching

The application of scientific research results in teaching directly enriches the teaching content, broadens students' horizons, deepens students' understanding of professional technology prospects, and enhances their professional identity. The vivid examples of the integration

Table 1. Scientific research examples are directly used in the list of teaching contents

No.	Transformation of scientific research achievements	Teaching application (course name, experiment name, etc.)	Application initiation time
1	Test bed of planetary wheel soil interaction characteristics	Integrated curriculum design II	2009 present
2	The mobile system of the six-wheel planetary rover with hybrid adaptive suspension	Course experiment (robot technology and application course)	2010 present
3	Folding and unfolding mechanism of spacecraft	Bar mechanism (mechanical principle course)	2003 present
4	Missile control mechanism	Separation mechanism (aircraft structure design)	2014 present
5	Spatial connection and separation mechanism	Low impact separation device (space intelligent structure and mechanism)	2016 present

of science and education show the students a rich and colourful technological world, allowing them to get in touch with practical cases from engineering practice.

The seemingly out-of-reach cutting-edge technology is interpreted by teachers in the form of self-presentation. In these cases, there are mechanical arms in space stations, folding and unfolding mechanisms in spacecraft, and metal rubber components in buffer devices of the lunar lander. These contents not only show the scientific research characteristics of the college 'based on aerospace and serving national defence', but also show the professional and strong scientific research strength and solid technical foundation. This content of promoting teaching through research and integrating science and education has been widely praised by students.

4.3 School-enterprise linkage practice and practice

Production practice and graduation practice are important to practice links of college students' training. In recent years, due to the continuous expansion of enrolment scale, taking the Institute of Mechanical and Electrical Engineering as an example, the number of interns reaches 3–400 every year, all of whom go to Changchun FAW Factory for internships. Due to the limitation of internship enterprises' acceptance ability, the original participatory internship has become a visiting internship, and the effect of going out for an internship is getting worse and worse. For the sake of production management and confidentiality management, internship enterprises are generally unwilling to accept college students' internships and provide support and assistance for internships. The aircraft manufacturing engineering major of Harbin Institute of Technology established the orientation of individualized, elite personnel training, and put forward the decentralized internship method for the first time in the internship session. Students majoring in aircraft manufacturing engineering were dispersed to several aerospace enterprises for internships, and new internship bases such as Beijing Satellite Manufacturing Factory, Beijing Aerospace Aircraft General Design Department, Tianjin Aerospace Electromechanical Research Institute, Hafei Group, etc. were opened. These bases are important national spacecraft and aircraft design and manufacturing bases, and professional students were grouped according to internship bases. Each group has no more than 20 people. We divide the students of an internship base into several groups according to the type of work, such as numerical control, inspection, assembly, surface treatment, etc. Distributing the internship students to workshops and work stations, and conducting specific operations under the guidance of the master, and rotating according to the type of work every week to ensure that every student can get exercises of various types of work.

Figure 6. Students' Internship in Tianjin Aerospace Electromechanical Research Institute

Figure 7. Students' Internship in Aerospace Test Base

To enable postgraduates majoring in aircraft manufacturing engineering to get more engineering training and exercise opportunities, and promote technological innovation, school-enterprise scientific research cooperation, and transformation of technological achievements. Relying on the decades-long scientific research cooperation foundation between Harbin Institute of Technology and China Academy of Space Technology, in 2012, a joint training base for graduate students in mechanical engineering was established in China Academy of Space Technology.

The General Design Department of Beijing Spacecraft, a subsidiary of the Academy, is the overall spacecraft research and development unit with the largest overall spacecraft field and the most complete professional technology in China. Beijing Satellite Factory is a high-tech enterprise specializing in the research and production of spacecraft. The two units represent the highest level of domestic spacecraft design, R&D, and manufacturing. In 2017, the base was selected as the national joint training demonstration base for graduate students with professional degrees. The Institute of Mechanical and Electrical Engineering of Harbin Institute of Technology sends 20–30 graduate students to the joint training base every year to engage in internship practice activities, and the internship practice time is 6–10 months.

Figure 8. Students' carry out practical activities in the off-campus joint training base

The joint training graduate students have participated in the research work of model product research and development and forward-looking basic pre-research projects jointly undertaken by both parties with high quality. The research topics include space manipulator, lunar sampling, large-scale deployable antenna, spacecraft connection, and separation, etc. In terms of training process management, both schools and enterprises jointly complete the teaching and management of graduate students stationed in the joint training base using 'sectional responsibility and collaborative management', and establish a joint training mode of dual tutor system for graduate students. Through the construction of a joint training base, the comprehensive quality and engineering practice ability of joint training graduate students have been greatly improved. Zhengwang Yu, a graduate student, won the 'Fifth National Master of Engineering Internship Excellent Achievement Award'. In addition, the employment competitiveness and career development potential of graduates are superior to the traditional training model, which further promotes the deep cooperation of scientific research between schools and enterprises, and improves the basic and innovative research level of enterprises.

4.4 Off-campus expert courses and innovative course construction

To broaden student's academic horizons and understand the development frontier of this major at home and abroad, we will jointly build advanced high-level undergraduate courses with domestic industry experts and well-known foreign scholars.

From the General Department of the Fifth Academy of Aerospace Engineering, the researcher Jianzhong Yang, an expert in the lunar exploration engineering structure institution, was hired to teach the spacecraft mechanism technology course. This course tells the basic contents of spacecraft mechanism and its design, spacecraft mechanism design and production process quality management, simulation analysis of spacecraft mechanism, and ground test of spacecraft mechanism from the perspective of engineering design and practice. Jianzhong Yang, based on his research and development experience of the landing buffer mechanism of Chang "e-IV", and closely combined with the engineering background of the space mission, taught the students the knowledge of the overall development of spacecraft, the research and development of key components, the ground test and the overall test verification, and introduced the common simulation methods of space flight and the future space simulation methods, which laid a good foundation for students to carry out aerospace engineering and technical work in the future.

Prof. Hornsen Tzou, an expert in structural electronics from the University of Kentucky, USA, taught the undergraduate students about the course of intelligent structure and intelligent control. This course introduced the mechanical basis of an intelligent structure, set up topics around the spatial application of intelligent structure, and students were divided into groups to research topics, gave design schemes, wrote analysis programs, and made physical prototypes. Organizing students to answer the final questions, and combining theoretical knowledge with solving practical problems through open course teaching. Cultivated students' innovative thinking in scientific research and ability to solve engineering problems.

The discipline gives full play to the advantages of aerospace engineering research, offers innovative training courses with aerospace characteristics for senior students, and offers 8 innovative training courses around the cutting-edge directions of planetary movement detection, planetary sample collection, large-scale space development, additive manufacturing, and space robots. The innovative practice was carried out on weekends in the laboratory with small class flexible teaching of 5–15 students, and we extract training objects from scientific research projects, clarified course contents, and objectives. The innovative work is carried out in groups, so that every student can have the opportunity to participate in the work of scheme design, modelling analysis, material production, etc., and evaluate the students' completion through the final reply. Training students' ability to solve practical problems by using what they have learned, and then laying a scientific research foundation for graduate students to study and participate in work.

5 Achievements in talent training

In recent years, remarkable achievements have been made in the cultivation of professional talents with students as the centre, research and teaching as the way, and personalized cultivation as the mainline. After application and practice in the aircraft manufacturing engineering major of HIT, remarkable results have been achieved.

5.1 Theoretical research of personalized training system

In 2017, "Construction and Implementation of Innovative Practice System of Combining Reality with Reality for High-end Equipment Manufacturing Talents Training" and "Research and Practice of Graphics Curriculum System for New Engineering Talents Training" won the second prize of teaching achievement in Heilongjiang Province respectively. In 2018, "Construction of Basic Course System of Mechanical Design for New Engineering Based on Ability Orientation" and 'Exploration and Practice of Training Mode for New Engineering Talents of Mechanical Major Oriented to Intelligent Manufacturing' were approved by the Ministry of Education for New Engineering.

5.2 Teachers and students' innovative ability training

The effectiveness of research feedback teaching has gradually become prominent, and the interaction between teachers and students has promoted the improvement of professional teachers' scientific research ability and comprehensive level, also significantly improved students' innovative practice ability. In 2016, the virtual simulation team of high-end equipment manufacturing was selected as the research-oriented teaching innovation team of the Ministry of Industry and Information Technology. In 2018, the aerospace agency and control team were rated as the national key field innovation team. In recent 3 years, students of this major have won 32 national and provincial competition awards, including the Gold Award in the International Vertex Design Competition, the National University Student robot contest Special Prize and other awards representing the highest level of domestic undergraduates.

5.3 Improving the quality of graduates

The quality of personnel training is highly praised and favoured by domestic and foreign universities, domestic research institutes, aerospace enterprises, and well-known private enterprises. A sample survey of 123 undergraduate graduates of aircraft manufacturing engineering in recent three years shows that there are currently 51 graduates studying for master's degree in domestic universities, accounting for 41.5%, all of whom have achieved excellent results during their master's degree studies. 13 graduates are studying for master's degree abroad, accounting for 10.6%, among whom 5 have studied for the doctoral degree. Graduates are engaged in product development, scientific research, and management in the fields of aircraft, robotics, vehicle engineering, information, nanotechnology, etc., and the employment rate of graduates reaches 100%. In recent years, 59 people have worked in scientific research institutes, aerospace enterprises, Sino-foreign joint ventures, and government aircraft, among them, nearly 10 people have served as project managers and business backbones in aerospace enterprises, and some graduates have become backbone members of major national aerospace projects and key models of new generation fighters.

6 Conclusion and prospect

Aiming at building the first-class major of aircraft manufacturing engineering, we have strengthened the comprehensive major reforming idea of "personalized talent training", established the concept of personalized talent training integrated with new aerospace technologies, popularized and applied the teaching method of "research-based learning", and constantly strengthened the training objectives for aerospace, serving national defence and aiming at major national equipment. We have also optimized the professional knowledge system, built high-level courses with overseas scholars and domestic industry experts, set up an off-campus practice training base for personalized talent training, fed back teaching with scientific research, strengthened the cultivation of practice and innovation ability, and taken various measures to promote comprehensive professional reform, thus steadily improved the quality of talent training.

In the future, we will continue to improve and deepen the personalized personnel training method system, seize the historical opportunity of HIT to build a national laboratory of aerospace technology, further integrate the latest aerospace technology into the teaching and training system, and put aerospace technology into the whole chain of personnel training of teaching-scientific research-practice-entrepreneurship. We will strengthen the joint training of talents with similar international universities and majors, and establish the "2 + 3" talent training mechanism. After studying basic courses for two years in China, students will go abroad for three years of professional course study and practical training. Mutual recognition of courses and credits, exchanges of students and teachers will be realized. After graduation, students will be awarded diplomas and degree certificates from Harbin Institute of Technology and foreign universities, respectively. We will continue to explore the new curriculum system and training mode of new aerospace technology, and cultivate a new generation of aircraft manufacturing professionals with international vision and pioneering spirit.

References

[1] Z. Pang, *Chang'e flying to the moon—China's first lunar envoy*, Knowledge is power, **1**, 10 (2014) (in Chinese)

[2] URL `https://m.xyxun.com/section-3-201254-0.html`

[3] X. Li, L. Qiang, H. Aihua, H. Fang, *Individualized Talent Training Mode Based on the "Stratified and Classified" Curriculum System*, Research in Higher Education of Engineering, **4**, 124 (2021) (in Chinese)

[4] S. Jiazeng, W. Weichao, Z. Weiping, *Exploration and informatization practice of undergraduate tutor system oriented to personalized training*, Journal of Architectural Education in Institutions of Higher Learning, **28(4)**, 34 (2019) (in Chinese)

[5] H. Shunshi, Z. Liang, W. Tao, Q. Jianxin, *Exploration and practice of personalized talent training and first-class specialty construction*, Journal of Higher Education, **25**, 15 (2021) (in Chinese)

[6] S. Shiping, L. Qiwang, H. Zheng, *Exploration and Practice of Aerospace Manufacturing Talents Training Mode Based on Integration Between Industry and Education*, Journal of Nanchang Hangkong University, **35(1)**, 109 (2021) (in Chinese)

[7] S. Lei, H. Jun, *Research and Practice on teaching reform of aircraft design specialty for the cultivation of innovative talents*, Academic Degrees & Graduate Education, **3**, 47 (2011) (In Chinese)

[8] G. Xingwu, W. Huarong, *Research on the evaluation index system of the construction of industry university research joint graduate training base*, Modern Education Management, **7**, 53 (2011) (in Chinese)

[9] X. Suli, X. Jiayu, P. Chaoyi, Z. Jiangcheng, *An Exploration and Practice of the Base Construction of Cultivating the Graduate's Innovative Ability through the Co-operation of the University and Enterprises*, Journal of Higher Education Research, **35(4)**, 107 (2012) (in Chinese)

[10] R. Subo, Z. Xingzhong, J. Xin et al., *Practical teaching exploration based on engineering education in machinery equipment major*, Research in Teaching, **39(4)**, 105 (2016) (in Chinese)

On training personnel for the interregional research and manufacturing cluster of aviation industry enterprises

Marina Nikandrova[1,*], *Roman* Moiseev[1], and *Alexander* Snegurenko[1]

[1]Kazan National Research Technical University named after A.N. Tupolev, 10, K. Marx st., Kazan, 420111, Russia

Abstract. The purpose of the article is to develop proposals for improving the efficiency of the forward-looking training system for the Russian Aviation Industry, located in the Volga Federal District. This territory is a concentration of enterprises of the machine-building complex, a significant part of which is represented by enterprises of the aviation industry. It is proposed to consider the aviation industry of the Volga Federal District as an interregional research and manufacturing cluster. The author presentation of the element composition of this cluster includes a production base, infrastructure, a system of state control, as well as consumers. As the core of this cluster, it is proposed to consider research and academic organizations, and industrial enterprises of the participant regions. Tatarstan is central to the Volga Federal District. This makes it a fairly convenient region in terms of logistics and interregional management of aviation industry enterprises. As the central headquarters of the interregional research and manufacturing cluster of the aviation industry, there is a scientific and educational institution in which scientific activity in the field of design and production of aviation equipment is set at a high level, as well as an extremely important function is implemented, namely the training of highly qualified personnel for the industry. The cases of KNRTU–KAI in the field of career guidance, teaching talented students, as well as the implementation of master's degree programs are considered. Many universities have got number of other similar unique practices for organizing staffing for the future development of enterprises in the aviation industry. In the authors' view, the significant potential for the development of the training system for the aviation industry lies in the efficient interaction between university communities within the Volga Federal District, as well as universities and employers. This will ensure consistency and continuity in training.

The Volga Federal District is a region that has concentrated a large number of enterprises of the aviation industry: in the Republic of Tatarstan there is a significant number of such production enterprises and scientific organizations: Kazan Aviation Plant—a branch of Tupolev PJSC, Kazan Helicopter Plant PJSC, MVEN Firm LLC; Tulpar Aero Group Group of Companies; Kazan National Research Technical University named after A.N. Tupolev—KAI; JSC Kazan Research Institute of Aviation Technologies, and many other subjects of the aviation industry; in the Republic of Bashkortostan, an aircraft manufacturing enterprise operates in the city of Kumertau, and aircraft engine production is carried out in the city of Ufa; in the city of Perm, the production of aircraft engines and components for them is carried

*e-mail: MVNikandrova@kai.ru

out (Perm Engine Company, JSC Perm Aggregate Association "Inkar", JSC "Aviadvigatel", JSC "Perm Motors"); the world's largest aviation-industrial complex Aviastar operates in the Ulyanovsk region, where the serial production of wide-body heavy and super-heavy transport and passenger aircraft of the AN-124 Ruslan and Tu-204 types is established; in the Samara region there is an aviation enterprise JSC "Aviakor", which produces Tu-14 aircraft, space techniques such as space carriers "Proton", carrier "Energia", rocket and space complexes "Soyuz-Progress", spacecraft "Foton", "Bion", "Resource"; in the Saratov region, the production of Yak-42 aircraft with various modifications is located, the production of multipurpose sports aircraft is also carried out. Due to the fact that these enterprises are sophisticated research and production complexes, they currently remain relatively closed to the external environment. Considerable reserves for increasing the efficiency of scientific and technical activities and production results of these enterprises lie in the organization of their coordinated and mutually beneficial interaction. The solution of this problem is directly related to the improvement of the system of forward-looking training of highly qualified personnel. Taking into account the development of production technologies and the spread of *Industry 4.0* technologies, to which the high-tech aviation industry is most susceptible, it is important to train personnel taking into account the results of the analysis of the experience of automating production processes and improving the tools of human-machine interaction [1]. At the same time, the pace of development of the modern aviation industry requires the organization of effective training of specialists at all stages of their professional life cycle, knowing as life-long learning. In this regard, the most urgent are the issues of effective retraining of personnel in the framework of continuing vocational education [2].

Today, in almost all regions of the district, training is organized for various levels of education—vocational and higher education. At the same time, work on training highly qualified personnel has been effectively organized in scientific centers and technical universities. However, the system of training highly qualified personnel for the industry has significant development potential. The training of personnel for the aviation industry, according to the authors, should be carried out in the interaction of specialized departments and organizations in the regions interested in the future development of the Russian aviation industry.

In this regard, the purpose of the article is to develop proposals for improving the efficiency of the forward-looking training system for the Russian aviation industry, in particular, localized in the Volga Federal District.

In the view of the authors, the concept of Aviation Industry is very broad. The number of enterprises and organizations of the aviation industry should include all economic entities associated with the design, production and operation of aviation equipment, as well as all entities that ensure the functioning of the relevant infrastructure, e.g. research, training, financing, insurance, etc.

Currently, there is a growing demand for aviation equipment on the Russian market. This is confirmed by the state program of Russia "Development of the Aviation Industry" adopted on September 21, 2021, in which the goal is to increase by 2030 the share of Russian-made aircrafts and helicopters in the fleet of the largest Russian air carriers at least 30 and 90%, respectively. So, according to this state program, it is planned in 2022 to have 14% of the fleet of domestically produced aircraft within the fleets of Russian air carriers (amount of 14% is the target for the coming year). Accordingly, every year the state plans to increase production volumes and the number of operated Russian aircraft, including on domestic route lines.

Along with the program documents, the growing trends in demand for air transportation and, as a result, the prospective demand for aviation equipment are confirmed by the data of the SberPress service. According to this service, the Russian aviation industry is recovering at a high rate, the volume of domestic air transportation by the end of 2021 will reach

81.6 million people, and will exceed the 2019 pre-COVID indicator by 12%. It should be acknowledged that the pandemic has become a new factor in socio-economic development. Due to epidemiological limitations, the aviation industry turned out to be the most sensitive to this fact [3].

Achieving the relevant indicators and meeting the demand for air transportation requires, first of all, an increase in the efficiency of forward-looking training for the aviation industry. In the authors' view, the key task is to ensure the efficiency of the functioning of the industry infrastructure, in particular, personnel training, as well as financial and economic support for the growth of aircraft production and economic conversion [4]. In this regard, the management and organizational problems of the development of the aviation industry of the Russian Federation in modern conditions are being mainstreamed. New integration mechanisms are required to improve the efficiency of interaction between participants of aircraft production [5].

Figure 1. Volga Federal District as the territory of the interregional research and production cluster of the aviation industry

In modern regional management, the cluster approach has proven itself well. The essence of the approach is in the systemic consideration of homogeneous independent elements of the regional economy having individual properties. In this regard, it is proposed to consider the aviation industry of the Volga Federal District as an interregional research and production cluster. As the core of this cluster, it is proposed to define scientific and educational organizations and industrial enterprises of the participant regions.

Tatarstan is central to the Volga Federal District. This makes it a fairly convenient region in terms of logistics and interregional management of aviation industry enterprises (figure 1).

According to the authors, a scientific and educational institution should act as the central headquarters of the interregional research and production cluster of the aviation industry, in which scientific activities in the design and production of aviation equipment are carried out at a high level, as well as an extremely important function is implemented, namely the training of highly qualified personnel for the industry.

The most suitable organization, in accordance with the above parameters, according to the authors, is Kazan National Research Technical University named after A.N. Tupolev—KAI (hereinafter—KNRTU–KAI). It is proposed to view KNRTU–KAI as a specialized organization within the framework of the cluster being created. The organization occupies a central place in the system of the interregional aviation industry, has significant opportunities to attract various resources available through the implementation of its own organizational potential. We believe that in order to attract the resources of KNRTU–KAI, it is necessary to initiate the creation of an interregional research and production cluster of the aviation industry. It is envisaged that KNRTU–KAI will act as a specialized organization—the core of a new structure coordinating regional system-engineering projects aimed at developing the Russian aviation industry as a whole. It should be noted that KNRTU–KAI has effectively organized international cooperation within the framework of the German-Russian Institute of Advanced Technologies (GRIAT KAI), which makes it possible to ensure the active participation of students in international projects in a multicultural environment, and to involve teaching staff and industrial technologies for the development of enterprises of the aviation industry [6].

Among the important results of the implementation of the directions for the development of the interregional research and production aviation industry are the following: resource and legal support for innovative activities of KNRTU–KAI by other enterprises of the aviation industry; forward-looking training of personnel to achieve the goals and objectives of the regional research and production cluster of the aviation industry; information support and ensuring the possibility of effective communication; the use of public institutions in solving economic, social and environmental problems of the aviation industry [7].

The author's presentation of the composition of the elements of the interregional research and production cluster of the aviation industry is shown in figure 2.

The objects of management of the interregional research and production cluster of the aviation industry include: Infrastructure (Research Institutes, Design Bureaus, Universities, Aviation Training Centers, Financial Institutions, Airports, etc.); Production Base (Aviation Equipment, Aircrafts, Helicopters, Sport Aviation, Spacecraft, Engines, etc.), Consumers (Civilian Consumers, Military-Industrial Complex, Sport Aviation Enthusiasts); the System of State Control for the Aviation Industry (Federal Agency for Air Transport "Rosaviatsiya", Federal Aviation Regulations, State Support, etc.).

According to the authors, the main task of the interregional research and production cluster of the aviation industry is to create conditions for the long-term technological leadership of the Russian aviation industry based on the development of scientific, technical, production and personnel potential of specialized enterprises. The development of corporate culture at enterprises that is part of the interregional research and production cluster, according to the authors, should be based on the best properties of the Russian model of production management and the use of foreign practices with their adaptation to the Russian conditions of functioning of enterprises of the aviation industry [8].

As part of the organizational development of the cluster, according to the authors, it is necessary the following: transformation of the existing specialized organization of the cluster into the format of a project team; development of regulations on the activities of a specialized cluster organization; introduction of payment of membership fees of participants for the development of a cluster system of interaction; organization of monitoring of satisfaction

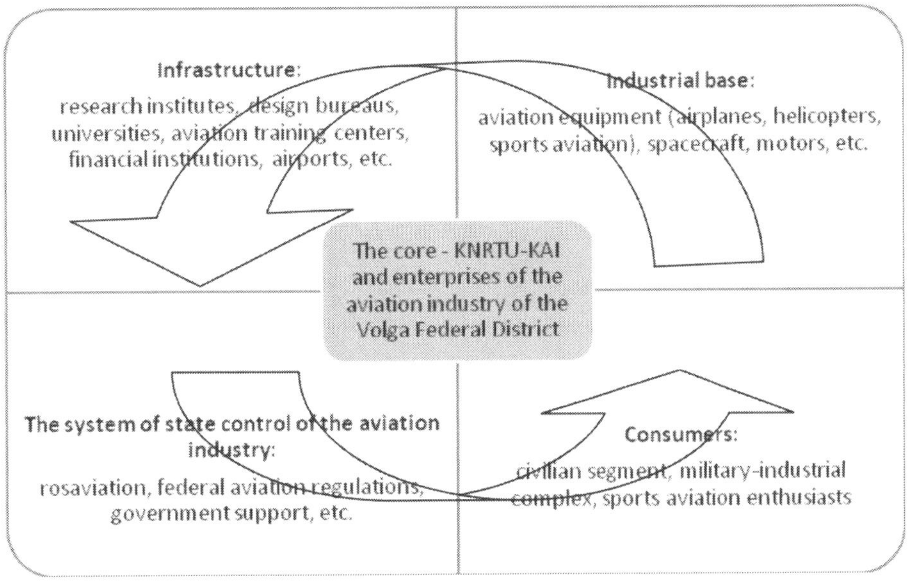

Figure 2. Author's presentation of the composition of the elements of the interregional research and production cluster of the aviation industry

with the activities of a specialized organization; creation of a technology commercialization center; creation of a prototyping center.

According to the logic of the National Technology Initiative, state development institutions should act as services that provide support, provide services to high-tech businesses. Companies need competent government services delivered on time and at a high level. In this regard, services should be created around the cluster that form a favorable breeding ground for cluster members. First of all, personnel training should become such a service.

KNRTU–KAI has been successfully solving the problem of training highly qualified personnel for the aviation industry for many years. This is a university with a tradition that has developed for almost a century. Of interest is the university's system of organizing career guidance aimed at identifying talents and developing them in the field of engineering, starting from an early age. So, KNRTU–KAI is implementing a project to create a unique educational complex "Technopolis" in Russia. This is an interactive space that integrates various forms of supplemenatry education for children in the field of physical and mathematical sciences, engineering and technical creativity. The complex includes three structural elements united by a single concept: the world's only interactive educational center for aviation and technology based on the first supersonic passenger aircraft Tu-144; engineering lyceum-boarding school for gifted children; children's scientific and practical laboratory located in the city center in a park area. The educational complex is a place for the development of children's abilities for technical creativity and entrepreneurship. At the same time, the concept of the educational complex provides for the consideration of the aviation industry, not only the production of aircrafts and helicopters, but small aircraft objects (drones and UAVs). At the same time, in order to identify children's abilities for technical creativity, KNRTU–KAI together with the Kazan Aviation Plant, is holding a profile Olympiad for schoolchildren in Physics. These and many other measures make it possible to ensure the identification and development of children's abilities for technical creativity in the field of aircraft construction and the operation of aircraft equipment.

Along with career guidance, KNRTU–KAI offers ample opportunities for the students who have already decided on their future profession and entered the university. Thus, KNRTU–KAI takes part in the All-Russian program for training engineers of a new generation for the construction industry "Wings of Rostec". The program is being implemented on the initiative and with funding from the State Corporation Rostec. In addition to the Kazan Technical University KNRTU–KAI, universities from other technical universities located in other cities of the Volga Federal District: Ufa, Samara, Perm, participate in the program. The essence of the project lies in the fact that in 2021, 120 applicants were recruited to Russian technical universities, who began their training in enhanced programs. It has been envisaged that from the very beginning of the training program participants will be given the opportunity to work in the structures of the State Corporation. Taking into account academic performance, all students receive motivational payments. The focus is on young people who are passionate about technical creativity and in the future will be able to ensure the development, including of enterprises of the aviation industry.

Graduates who have mastered bachelor's programs both at KNRTU–KAI and at other universities can enroll in international master's degree programs implemented at KNRTU–KAI. In order to effectively organize the educational process within the framework of the implementation of international master's degree programs at KNRTU–KAI, the German-Russian Institute of Advanced Technologies (GRIAT) has been created. This educational subdivision of KNRTU–KAI is a unique educational platform that unites German and Russian universities and enterprises. GRIAT trains engineers with global engineering skills and knowledge of a foreign language, develops international research and development, and also promotes intercultural exchange between Germany and Russia. Within the framework of GRIAT, various training programs are being implemented, including training of engineers in the field of aircraft engineering. Along with the training of aviators, GRIAT prepares masters in various fields of study. The competencies formed in master's students are in demand at enterprises of the aviation industry, as well as other industrial enterprises. The training activities are carried out at KNRTU–KAI for the aviation industry enterprises, and aimed at providing the industry with highly qualified specialists.

Thus, the aviation industry, as one of the most high-tech sectors, is undergoing major transformations today. In the face of constantly accelerating competition, the development of the aviation industry requires significant investments, which should bring maximum results. The article discusses the issue of creating an interregional research and production cluster of the aviation industry, in which KNRTU–KAI is considered as a specialized organization coordinating cluster projects. Tatarstan is central to the Volga Federal District. This makes it a fairly convenient region in terms of logistics and interregional management of aviation industry enterprises. The strategic areas of development of the interregional research and production cluster of the aviation industry include the following: organizational development of the cluster; development of the scientific and educational potential of the cluster; development of the information environment. This will make it possible to effectively attract and carry out joint investments in the system-engineering projects of the aviation industry. Cases of KNRTU–KAI in the field of career guidance, teaching talented students, as well as the implementation of master's degree programs are considered. Many universities have got many other similar unique practices for organizing staffing for the future development of enterprises in the aviation industry. In the authors' view, the significant potential for the development of the training system for the aviation industry lies in the effective organization of interaction between university communities within the Volga Federal District, as well as universities with employers. This will ensure consistency and continuity in training.

References

[1] L. Mládková, *Industry 4.0: Human-technology interaction: experience learned from the aviation industry*, in Proceedings of the 19th European Conference on Knowledge Management (ECKM 2018), 6–7 September 2018, University of Padua, Italy, **1**, 571 (2018)

[2] R.R. Anamova, L.V. Bykov, D.A. Kozorez, *Aircraft industry staff retraining as a part of vocational education in the Russian Federation*, TEM J., **8(3)**, 978 (2019)

[3] A.A. Sozinova, E.V. Sofiina, M.F. Safargaliyev, A.V. Varlamov, *Pandemic as a New Factor in Sustainable Economic Development in 2020: Scientific Analytics and Management Prospects*, Lecture Notes in Networks and Systems, **198**, 756 (2021)

[4] V.M. Kraev, A.I. Tikhonov, S.V. Novikov, *Economic conversion in the aviation industry*, Russian Engineering Research, **38(4)**, 330 (2018)

[5] A. Emir, *Urgent managerial and organizational issues in the industry of aviation in the Russian Federation*, in Theory and Practice of Management: the Response to Challenges of Digital Economy: Proceedings of X Scientific Conference for young scientists, 6 December 2019, Plekhanov Russian University of Economics, Moscow, Russia, 439 (2019)

[6] T.M. Tregubova, V.E. Kozlov, A.S. Kats, *Discource on students participation in international projects on inter-ethnic tolerance formation in multicultural environment*, Journal of Siberian Federal University. Humanities & Social Sciences, **12(2)**, 261 (2019)

[7] L.F. Henriksen, S. Ponte, *Public orchestration, social networks, and transnational environmental governance: lessons from the aviation industry*, Regul. Gov., **12(1)**, 23 (2018)

[8] M. Kulapov, Y. Odegov, V. Sidorova, N. Sidorov, E. Zotova, *Corporate culture of organization: typical and Russian model*, Montenegrin J. Econ., **15(1)**, 215 (2019)

Career guidance for the prospective students of KNRTU–KAI

Anna Nikolaeva[1,*], *Ekaterina* Zagrebina[1], and *Alexander* Snegurenko[1]

[1]Kazan National Research Technical University named after A.N. Tupolev, 10, K. Marx st., Kazan, 420111, Russia

Abstract. The article presents the experience of Kazan National Research Technical University named after A.N. Tupolev—KAI (the former HEI's name is KAI—Kazan Aviation Institute) in the implementation of the *policy* of career guidance of students and prospective students. The purpose of the article is to share the original KAI practices in the field of career guidance of prospective students, as well as to reveal the essence of the solutions incorporated into the key career guidance activities of the university. Original solutions are described, including: installation of the first supersonic passenger aircraft TU-144 on the territory of the KAI campus; the first scientific-practical conference of young aviators "Aviation of the Future: Trends, Challenges and Opportunities", dedicated to the 50th anniversary of the first flight of the TU-144 aircraft; creation of an educational complex "Technopolis" for the development of children's engineering creativity; an online course for schoolchildren during the university's career guidance campaign "KAI Online School CYBER.TECH"; extended day groups for schoolchildren; the project for the training of new generation of engineers "Wings of Rostec" implemented with the support of the State Corporation Rostec, and some other projects. The factors of entry and retention of personnel in professional communities are identified and systematized, taking into account the professional life cycle of a specialist. It has been identified that in modern conditions, career guidance activities have begun to significantly influence the updating of educational content and modes of study at the university, which is confirmed by the presence of interactive and online KAI events. The university forms and develops an environment favorable for self-determination and self-realization of all participants in the educational process, while covering a variety of target groups.

In the recent past, our country was the undoubted technological leader in the world—they were the first to launch a satellite and a man into space, built the first nuclear power plant and the world's first supersonic passenger aircraft. Today, domestic enterprises, admittedly, concede leadership in many areas of aerospace science and production to other countries. The restoration of Russia as a world aerospace leader is possible only on condition of advanced training of highly qualified engineering personnel in specialized educational institutions of vocational and higher education.

This problem is solved by the Kazan National Research Technical University named after A.N. Tupolev—KAI (formerly KAI, Kazan Aviation Institute) which is one of the leaders in Russian engineering education. Like any modern Russian university, KAI functions in the context of the growing competition for talented applicants and prospective students. The

*e-mail: Annanikolaeva@bk.ru

university gives priority attention to the quality of admission. The grade point average of those admitted to the first year is steadily growing: in 2016 GPA was 70.5, and in 2021 GPA was 76.1. In addition, the number of winners and prize-winners of the All-Russian Olympiads entering KAI is rapidly growing. Over the past 5 years, KAI has received 76 winners and prize-winners of the All-Russian and Federal Olympiads, that is the most of all among the universities located in the Republic of Tatarstan.

Nevertheless, it is worth noting that there are serious problems associated with the career guidance of potential applicants entering the engineering bachelor's degree and specialist's degree programs, which also matter for a large number of enterprises of the aviation industry in Russia [1]. These problems include the following:

1. A direct indicator of children's interest in technical occupations is their choice for taking the Unified State Examination in Physics. Unfortunately, the number of school leavers choosing this subject as their final exam remains relatively low. In recent years, there has been a tendency for an increase in the number of school leavers choosing humanities as an elective discipline to pass the Unified State Exam.

2. The quality of knowledge in this subject (Physics) leaves much to be desired: in recent years, the results of the Unified State Exam in Physics are lower than the ones in Social Sciences and Humanities.

3. The following fact is even more alarming. Schoolchildren in Tatarstan and Kazan show the best results in the country in most subjects. Physics, however, is an exception. The results of the average USE score in 2021 in Physics in Tatarstan amounted to 59.39, which is higher than the average in Russia, but far from the highest level among the Russian regions [2].

These problems acute the need to organize effective career guidance and counseling of prospective students based on early identification of their talents. According to the authors, special attention should be paid to the timely identification of the propensity for engineering and technical creativity in children. KAI has got a variety of original practices for implementation of career guidance for prospective students. This experience, according to the authors, can be extremely useful for other universities.

The purpose of the article is to share the original KAI practices in the field of career guidance for prospective students, as well as to reveal the essence of the solutions incorporated into the key career guidance activities of the university.

One of KAI's striking projects in the field of popularization of engineering and high-skills professions among children and youth, their patriotic education based on the greatest achievement of domestic designers is the installation on the KAI campus of the first supersonic passenger aircraft TU-144. The history of the creation of the TU-144 is the history of the great victory of the Soviet Union engineering school, since the design of the aircraft began 2 years later than the British-French Concorde, and the flight took place 2 months earlier. The world's first supersonic passenger aircraft has incorporated hundreds of world-class innovations. The university has carried out a complex operation to transport a 90-ton aircraft to the university campus. At the moment, work has been carried out on the external restoration of the side and its illumination. The plane at the university is the subject of patriotic education, and awakens a sense of pride in the achievements of domestic engineers of the 20th century, and demonstrates the capabilities of human potential.

In 2019, on the basis of KAI, the first scientific-practical conference of young aviators "Aviation of the Future: Trends, Challenges and Opportunities" was held. It was dedicated to the 50th anniversary of the TU-144 first flight, which took place on December 31, 1968 [3].

The main emphasis of the conference was made on the creation of opportunities for an open dialogue between students of aerospace universities and young personnel of aviation enterprises, as well as students of engineering classes at educational schools in Russia. All conference participants had the opportunity to speak at the poster session. The conference was aimed at bright and compelling popularization of engineering education among young people. The key results of the conference were the following: assistance to the patriotic education of children and youth based on one of the brightest symbols of the great Soviet engineering school; increasing the interest of children and youth in the natural sciences and technical creativity; strengthening the international brand awareness of Russian engineering education; emergence of a new tourist symbol in the Volga Federal District and the development of a Russian tourist product based on the TU-144; development of urban infrastructure for children's informative leisure, etc.

Another KAI's initiative in the field of career guidance for prospective students is the "Technopolis" project. The educational complex "Technopolis" is an interactive space that integrates various forms of supplementary education for children in the field of physical and mathematical sciences, engineering and technical creativity. The educational complex creates and implements innovative programs of supplementary education of a technical focus, using modern equipment and education technologies, which contributes to solving the problems of early identification of children's inclinations to engineering creativity. "Technopolis" uses modern digital engagement methods. The main principle of the complex's activity is to show children that engineering area is "cool", to awaken in them the confidence that they can also create a new legend of world engineering.

Both of the above mentioned projects were initially supported by the Russian Presidential Grants Fund.

Long before the start of the pandemic, KAI organized work on the use of online technologies both in the educational process and in the career guidance for prospective students.

We live in an era of digitalization and global access to almost any information through the World Wide Web. The generation of modern children is not called the digital generation for nothing, since from an early age many of them grow and develop in constant interaction with gadgets, social networks, etc. The vast majority of students in modern schools use modern digital technologies in the learning process. Thus, communication channels and interaction formats with prospective students are changing in career guidance of universities.

To expand access to a larger number of prospective students, as well as to increase interest in engineering educational programs at KAI, an online course of KAI Online School CYBER.TECH has been implemented.

The online course for schoolchildren during the university's career guidance campaign the KAI Online School CYBER.TECH allows you to transform a learning mode from "knowledge transfer" into creating conditions under which the processes of professional determination of the schoolchild-applicant become possible.

The aim of the KAI Online School CYBER.TECH is to familiarize the participants with the best fields of study at KNRTU–KAI in order to form a sustainable interest in engineering professions, and to help them choose their future educational trajectory.

The KAI Online School CYBER.TECH creates a favorable environment for enhancing the cognitive activity of the prospective students, and organizes the conditions for the formation of individual educational trajectories.

The results of the online course of CYBER.TECH have become the expansion of the geographical boundaries of the career guidance work of the university and the popularization of engineering fields of study.

An important step of the university in the development of career guidance for prospective students is the involvement of the intellectual potential of students and teaching staff to prepare and instill an interest in engineering among schoolchildren who are potential prospective students.

One of the most interesting solutions of the university for organizing career guidance for future applicants is the project entitled the "KAI Academy—Mini Wings". These are online after-school groups for schoolchildren. Within such online groups, KAI's students provide counseling and intellectual support for schoolchildren for completing their homework. Using online technology, almost any schoolchild in Russia and abroad can receive competent consultations from KAI's students on various subjects of the school curriculum. The idea of the project is to involve the most talented students in the process of interacting with potential prospective students. As already noted, in KAI there is a fairly high average score of annually enrolled students, by the number of budget places, more than 1,000 applicants become KAI students. Students are a significant intellectual potential of the university. The project "KAI Academy—Mini Wings" provides effective use of this potential for career guidance for future applicants. Due to the fact that students are close enough in age to the schoolchildren being taught, the age barrier is removed, trusting and friendly communication is built, and schoolchildren' loyalty to engineering and the university is formed. Due to the growing demand, this project has moved from the category of student social design to the category of student social entrepreneurship. The uniqueness of the project lies in the fact that today in the Republic of Tatarstan the "KAI Academy—Mini Wings" as online after-school groups for schoolchildren is the only support center for schoolchildren in completing homework in ten school disciplines. The project develops the entrepreneurial abilities of students based on network forms of organizing intellectual work [4].

Since 2020, KAI has been a member of the "Wings of Rostec" project. The project is aimed at training a new generation of engineers with a set of advanced digital competencies, integrated into the international academic space, fully ready to work in modern conditions at the enterprises of the aircraft industry. The State Corporation Rostec organizes the recruitment of talented prospective students to the aerospace universities of Russia, provides material support for students participating in the project, provides practice-oriented training and subsequent employment at enterprises operating in the management loop of the State Corporation. In addition to KAI, universities from such cities as: Moscow, Samara, Ufa, Ulan-Ude, Perm, Rybinsk, Irkutsk and Novosibirsk take part in the project. The curriculum development is carried out in full compliance with Federal State Educational Standards, and also includes a practice-oriented training modules, for example, Fail-safe, Fatigue Strength, Acoustics, Life Support Systems for Aircraft, Testing of Aviation Equipment, Quality Control, Operation of Aviation Equipment, Industry 4.0, Industrial Design, Additive Technologies, etc.

KAI has a unique creative laboratory—the Glass Processing Science and Technology Center, fitted out with specialized equipment. On the basis of the center, informational activities and master classes are held for schoolchildren and students on the processing of glass waste into ethno-souvenir products. In the future, the issue of using online technologies to ensure the availability of information for everyone is being considered. The Glass Processing Center is the only center for scientific, technical and artistic creativity in the Republic of Tatarstan based on glass processing. The Center's activities are aimed at preserving historical memory, environmental education, historical education, scientific, technical and artistic creativity of children and youth [5]. The center works with children and adolescents from 12 to 16 years old in Kazan and Tatarstan, who want to engage in technical and artistic creativity in the field of glassmaking, who are interested in the prevention and solution of environmental problems.

KAI has implemented a project to create an interactive exposition "Tupolev Park" on the university campus. The art object is located in the center of the spiral staircase of the five-story building of the German-Russian Institute of Advanced Technologies (GRIAT KAI).

The exposition is a decorative dynamic structure that simulates the takeoff and ascent into the air of aircrafts of the TU family. Volumetric models of the aircraft turbine on the ceiling of the fifth floor and the propeller on the floor of the first floor are connected by thin strong cables, on which exact copies of 15 different aircraft models of the TU family are attached in a vertical spiral. The rotation of the interconnected turbine and propeller sets a smooth movement around the circumference of the aircrafts with the effect of takeoff and landing. Models of 15 aircrafts, which have been created at the design bureau of the world aviation genius Andrey Tupolev, are already in themselves a vivid example of the pinnacle of Russian engineering thought of the twentieth century. All aircraft models have been donated by Tupolev PJSC. Climbing the stairs, now everyone has the opportunity to get acquainted with the history of the creation of aircrafts of the TU family, and learn more about the personality of the legendary designer and manager A.N. Tupolev. A specially developed application through the virtual reality system allows you to study in detail information about the manufacturer, developer, the first flight of the aircraft, as well as about technical specifications (length, height, wingspan, wing area), mass properties and flight data. All this makes it possible to study the history and features of the legendary representatives of the domestic aircraft industry for military and civil purposes in a modern interactive format, for instance, such aircrafts as TU-160, TU-214, TU-22M, TU-330 and TU-204. The project is aimed at the patriotic education of children and youth based on the greatest achievements of domestic aircraft designers. The interactive exposition is an important element of supplementary education and patriotic education programs for schoolchildren and students, as well as an interactive information stand for classes in the framework of project-oriented thematic lessons, workshops, engineering summer schools and other events that take place at the university.

University career guidance activities cover various categories of young people, including children without parental care. So, in 2020, an interesting project was implemented at KAI aimed at developing the abilities for engineering creativity in the children of the orphanage. The KAI team organized and conducted an educational action—a master class for pupils of the Nurlat orphanage located in the Zelenodolsk region of the Republic of Tatarstan. The project was initiated by three organizations working within the structure of KNRTU–KAI: an endowment fund, a trade union organization, a council of university veterans with the support of the Vladimir Potanin charitable foundation. Representatives of the KAI donated to the Nurlat orphanage the means of technical equipment for online learning (web-cameras, headphones, etc.) purchased with the funds of a grant from the Vladimir Potanin Foundation, as well as kits for technical creativity in robotics LEGO Education. The implementation of the project was aimed at solving current problems (technical equipment for online training) and solving promising problems (providing methodological materials, personnel and means for technical creativity). The authors of the project believed that there were many talented children among the inmates of orphanages, including those in the field of engineering. It would be important to organize and conduct events in a timely manner and involve young people in engineering creativity.

An important KAI initiative in the field of career guidance among young people is a project aimed at attracting applicants from specialized vocational education institutions: colleges and technical schools. The fact is that this category of applicants, having already a certain set of competencies and work experience, have significant potential for developing their own abilities, and can achieve fairly high results. Since 2017, KAI has held an annual competition of graduation works of secondary vocational education entitled "Diamonds of Science". Theses of final qualification works in the form of a presentation have been accepted for the competition. Contest works have been evaluated in the following nominations: Professional Orientation; Scientific Perspective; Best Design; Originality of the Idea. Upon admission to the academic programs of bachelors and specialists, the diploma of the winner and the certificate of the participant of this competition make it possible to accrue additional points to the results of the Unified State Exam or the results of internal entrance examinations conducted by KNRTU–KAI as an individual achievement. Having the opportunity to take part in this competition increases the awareness of potential applicants about the rules of admission to the university and increases the opportunities for admission to the programs they are interested in.

Thus, in KAI, as in many universities, there are many activities aimed at career guidance for prospective students. Particular attention is paid to the issues of early detection of talents in children, and the development of their ability for engineering creativity. Every year, the set of activities is reviewed, new projects are created and implemented to organize more effective career guidance.

All the above measures for career guidance for prospective students and students have been developed taking into account the identified factors of entry and retention of personnel in professional communities (figure 1).

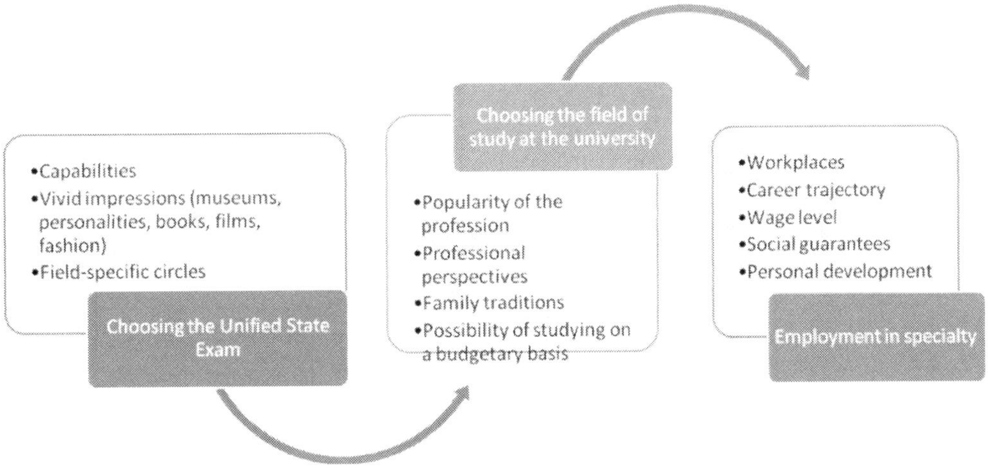

Figure 1. Factors of entry and retention of personnel in professional communities

The authors have identified the following stages: choosing the Unified State Exam, choosing the field of study at the university, obtaining employment in the specialty during the professional life cycle, and the factors corresponding to the stages.

At the stage of choosing the Unified State Exam, according to the authors, such factors as: individual abilities have a significant impact on the possibility of entering professional communities; vivid impressions (museums, personalities, books, films, fashion) received at an early age; the opportunity to visit field-specific circles and sections.

At the stage of choosing the field of study at the university, according to the authors, a significant impact on the possibility of entering and consolidating in professional communities have such factors as: popularity of the profession; professional and financial perspectives; presence of family traditions; possibility of studying on a budgetary basis, etc.

At the stage of employment of a trained specialist, according to the authors, the respective factors are: presence or absence of jobs in the industry; potential career trajectory; wage level; social guarantees; conditions for personal development, etc.

The university proceeds from the fact that during the professional cycle, each person goes through certain stages that have a significant impact on the possibility of entering and consolidating in a certain professional community. According to the authors, one cannot ignore the fact that, along with the competition between universities for talented prospective students, the competition between the direct fields of study is growing. Taking these factors into account makes it possible to form measures for the career guidance of professional communities' members at various stages of their professional life cycle, and implements educational programs for various age categories of students.

Recently, vocational guidance activities have begun to significantly influence the updating of educational content and modes of study at the university, which is confirmed by the presence of interactive and online KAI events. Any actions are aimed at increasing the integral indicator of the quality of university services [6].

Taking into account the accelerating dynamics of the demand for professional competencies presented by employers and directly by the students themselves, KAI forms and develops an environment favorable for self-determination and self-realization of all participants in the educational process, while covering various target groups of students.

References

[1] R.R. Anamova, L.V. Bykov, D.A. Kozorez, *Aircraft industry staff retraining as a part of vocational education in the Russian Federation*, TEM J., **8(3)**, 978 (2019)

[2] A. Svirina, A. Lopatin, J. Titko, *Analysis of students performance in relation to the results of State Unified Exam: The case of Russian university*, Business, Management and Economics Engineering, **19(1)**, 170 (2021)

[3] A.V. Kornilin, M.F. Safargaliev, *Economic Evaluation of Measures to Improve the Efficiency of Knowledge-Intensive Production in a Networked Economy*, in Proceedings of the 1st All-Russian Scientific-Practical Conference of Young Aviators of Russia "Aviation of the Future: Trends, Challenges and Opportunities", 29–30 November 2019, Kazan, Russia, 411 (2019)

[4] M.F. Safargaliev, Y.A. Kitsay, E.N. Egorova, L.V. Ermolina, *Modernization of modern entrepreneurship on the basis of artificial intelligence*, Lecture Notes in Networks and Systems, **91**, 346 (2020)

[5] T.M. Tregubova, V.E. Kozlov, A.S. Kats, *Discource on students participation in international projects on inter-ethnic tolerance formation in multicultural environment*, Journal of Siberian Federal University. Humanities & Social Sciences, **12(2)**, 261 (2019)

[6] M. Mironova, L. Nugumanova, A. Bushueva, S. Markova, Z. Zapparova, *Integrated indicator of quality of services of higher education institutions*, in 19th International Multidisciplinary Scientific GeoConference SGEM 2019, 30 June—6 July, 2019, Bulgaria, **19**, 191 (2019), DOI: 10.5593/sgem2019/5.4/S22.026

Methods of adaptation of first-year students in higher education institution, as a basis for involvement in academic and student life at an early stage of education

Denis Prokopenko[1,*] and *Vera* Kaigorodova[1]

[1]Department 101 "Design and Certification Aviation Technology", Moscow Aviation Institute (National Research University), 4, Volokolamskoe shosse, Moscow, 125993, Russia

Abstract. While studying at a higher educational institution, a student is immersed in a unique world with its own rules and laws. Training in various forms of specialization lasts from 4 to 5.5 years. During this time, the student manages to replenish his knowledge base with unique competencies and qualities that will later help to achieve successful career growth. It is obvious that getting a higher education plays a significant role in the formation of a personality, higher schools have everything necessary to test the acquired knowledge and skills. The flexibility of the curricula allows students to choose which disciplines they want to study. But before starting to function comfortably in an educational environment, students must go through adaptation. In the entire history of the existence of educational organizations, there was no higher educational institution where the student felt comfortable and cozy from the first days. This article pays special attention to the methods and forms of undergoing the adaptation period for first-year students on the example of the implementation of a number of activities taking place at the Institute No. 1 "Aviation Technology" of the Moscow Aviation Institute (National Research University).

1 Analysis of modern forms of adaptation of students at the university

During the entire existence of higher education schools, the image of the student is constantly changing and improving. If we turn to the last century and delve into the history of students, we will notice that the modern student is completely different from the student of the 20th century. Therefore, a logical question arises: "What is the reason for this?" Analyzing the history of the development of education, as well as the periods of time and situations that arose at certain points in history, we can conclude that education directly depends on the needs of a person and changes depending on his/her requests. Currently, the industry is on the verge of digital transformation, many enterprises are switching to internal management through digitalization [1]. This explains the high request of IT specialists, whose profession is considered to be one of the most popular in the modern world. The next scheme reflects student environment's dependence of the state's needs (figure 1).

The following follows from this scheme: the government, through its state-owned companies and departments, analyzing the situation in the country and monitoring the development of all spheres of consumption, interacts with higher educational institutions about ordering

*e-mail: prokopenkoda@mai.ru

specialists in demanded areas in the country. A higher educational institution, receiving an order from the state, carries out recruitment according to its main educational programs, forming a student environment for these programs and for itself. This can be concluded on the basis of the release of a huge number of specialists in the field of engineering and IT programming, although 10–15 years ago everything was radically different [2]. During that time the government required more specialists in the field of economics, finance, banking, etc., this was mainly due to the post-industrial period, when the state produced various types of products over the needed quantity and the requirement on the specialists in the field of sales arose to sell those products to the partners.

Figure 1. Student environment's dependence of the state's needs

A person who is a specialist in a specific field of activity has a certain set of professional competencies that are laid down in him at the stage of education in a higher educational institution, hereinafter referred to as a university. It is obvious that the competencies that should be decisive in a particular profession are different for each person, but they help to solve problems during professional activities.

At the stage of admission to the university, applicants have only personal competencies that were brought up in them from birth by their parents and the environment in which they grew up and developed. As we can guess, applicants do not have professional competencies at the moment, since they have not yet studied their future profession and do not know what professional qualities they will need. If we perceive such an area of development of professional competencies in a person, then the university occupies the main position in it, since during training, students, in addition to participating in seminars and lectures, have the opportunity to undergo practical training at enterprises or companies in which they after graduation from the university can get a job. It is practical trainings that enable future specialists to test their professional competencies in practice solving real tasks set by a future, potential employer.

Many people know that the education of students at the university is divided into several stages. We present the proposed stages in a step-by-step model of the existence of a potential student in an educational environment which if reflected on the figure (figure 2).

This diagram clearly shows what stages a student must go through in order to master professional competencies and obtain a diploma of higher education. Of course, all these steps are very important, and have the effect of sequential accumulation of knowledge—without overcoming the upcoming stage, a student cannot move to a new stage, since only a sequential algorithm for obtaining knowledge is the most effective in modern society. At the very beginning of the scheme, we see the first stage, where the applicant gets acquaintance with the university and overcomes the first serious stage of selection—entrance tests. The second stage shows us the student's admission to a higher educational institution and the beginning of the adaptation period. Our article is devoted to this period and methods of adaptation [3].

In this article, we will analyze what is meant by the phrase "Social adaptation of students at the university", what difficulties students face when getting into a new unexplored environment of a higher education school, what methods should be used to adapt students as soon as possible, as well as what activities or social conditions should be fulfilled in order for a student to feel "at home" in a new educational team.

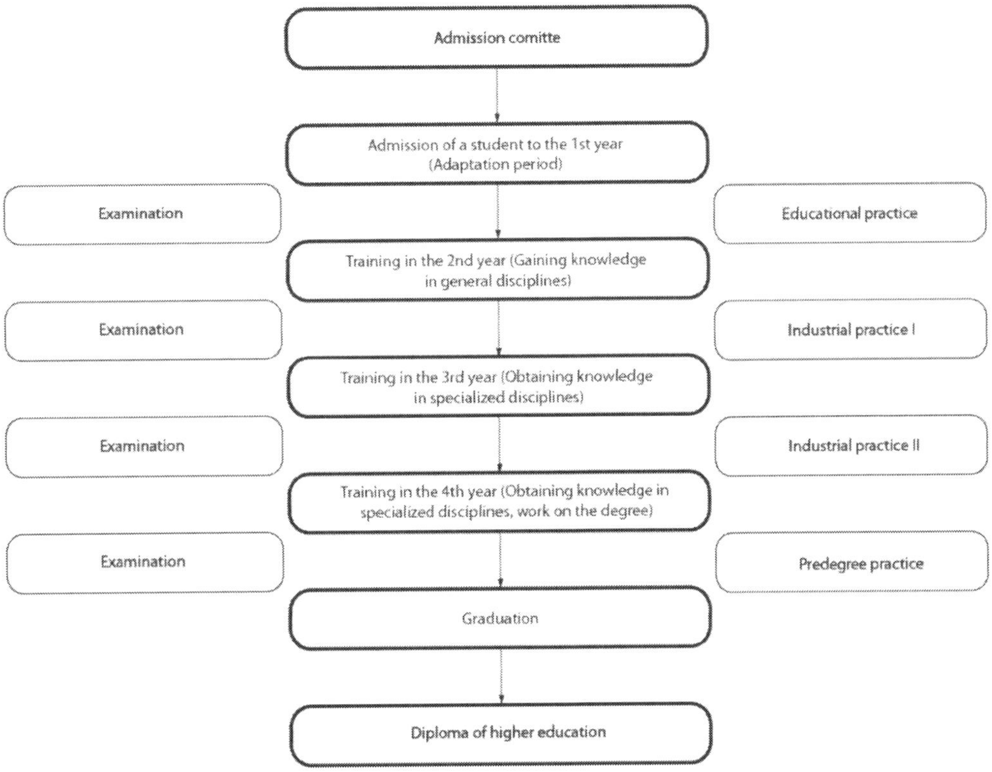

Figure 2. The path of a student while studying at a university

The social adaptation of a student in a higher educational institution is understood as the ability of a subject (student) to adapt to a new, unexplored environment, to find points of interaction with new subjects (faculty, classmates) for further comfortable existence in a single team. As a rule, students' social adaptation begins in the 1st year of study and lasts for a total of 1 year. Researchers in this field distinguish 3 forms of adaptation:

1) formal adaptation—scanning and studying the signs of behavior of objects in a new educational environment through social relationships;

2) social adaptation—integration of students and their further unification into a new collective (student group) through communication and study of the subject and integration of these study groups into the student environment;

3) didactic adaptation is the preparation of students for new forms of studying educational material, as well as their participation in student events.

A significant part of adaptive situations arise as a result of changes in learning conditions when entering a university. This aspect of adaptation is called didactic adaptation, which includes adaptation to new forms of control. The adaptation of first-year students is closely related to their academic success. The purpose of didactic adaptation is to develop optimal methods of organizing the educational activities of first-year students, taking into account the specifics of the learning environment and the peculiarities of cognitive activity at a young age. The success of students' education depends on many factors, one of the most important is their intellectual development as an indicator of mental activity.

Let's consider the adaptation period of students on the example of the Moscow Aviation Institute (National Research University), hereinafter—MAI (NIU).

As a rule, a new student collective (study group) unites a community of students of the same age. All of them are radically different with a set of individual achievements and personal qualities, but they are united by the goal of obtaining a higher education diploma. Naturally, students have a certain amount of excitement before the start of the academic year, which is caused by several factors:

1) a completely new team in which they must exist and find points of interaction;

2) unknown form and methods of education.

It is obvious that it is almost impossible to cope with the above-described factors and overcome such a difficult and thorny path alone, so students in the first days of training try to establish communication with the same freshmen as them, try to find comrades close to themselves in spirit and ideology. Of course, the concept of higher education is based on the acquisition by students of professional skills and knowledge for further professional field. The university is the next body after the school, which is engaged in the educational and upbringing activities of students to form the principles of ideological equality, the principles of humanism, behavior in society, and so on. The teaching staff of universities have always acted and are currently acting as mentors for the student community. Of course, such an important stage as social adaptation is not ignored, and senior colleagues and teachers constantly interact with first-year students to overcome the adaptation period as soon as possible [4].

Before proceeding to the description of the way to overcome adaptation, it is worth analyzing the problems and barriers faced by students who have opened the doors of the university for the first time.

Of course, one of the main reasons for concern is a significant increase in the volume of the academic load, the emergence of new unfamiliar disciplines, the development of which requires in-depth knowledge in related fields. Next comes a completely different and unknown format of the educational environment into which the student gets. The diversity begins, in which the student needs to exist. There are different formats of classes—seminars, lectures, practical works, independent study of methodological manuals and literature. In addition, student needs to have time to find time to participate in student life.

Another barrier that many first-year students face is the development of skills for practicing didactic material during training sessions. Preparing for seminars, taking notes of lectures, and performing laboratory work often require completely different skills and techniques that are absolutely incompatible with the school methodology. This is also something freshmen have to learn.

The next separate problem for first-year students is interaction with the teaching staff of the university. Higher school teachers often differ in their methodical approach to the disciplines taught, it is obvious that each teacher has his own personal method of teaching. Besides, teachers differ from school teachers in academic titles and degrees, which gives them an undeniable weight in the scientific community. All of the above descriptions can make first-year students uncomfortable during first meetings with teachers and be a significant "stumbling block" for interaction [5].

Of course, another determining factor in the anxiety of future students is the relationship with classmates. Building such relationships is not always easy and painless, because the formation of a new team of people is often accompanied by conflict situations, drawing attention to themselves, the distribution of "roles", which is not a very pleasant process.

Also, student life itself is not as unambiguous as it seems at first glance. Student life directly depends on the student's self-organization. As far as you have developed internal discipline and organization, so you will be open to the new world of students. In many universities, at the level of educational standards, there is no such factor as constant assessment of knowledge and attendance control, so many students have a feeling of complete freedom, as well as the temptation to use their time not from studying but on themselves.

2 Methods of adaptation of students at the university

Having analyzed the problems faced by first-year students at the initial stage of education, it can be concluded that the adaptation of students is a complex, dynamic and multilevel process of rethinking and restructuring the need-motivation base, the complex of existing personal competencies and skills in accordance with new goals, objectives, prospects and conditions for their implementation. Turning to the methods of social adaptation of students in a new educational environment, it is impossible to pass by the most basic and main method—unequal interaction with the applicant at the stage of submitting documents to the university.

The main source of anxiety among young people at the stage of joining a new team is the complete absence of personal contacts and familiar faces in the new team. In order to solve this problem, it is necessary to put together the image of the university as a place in which the student will be comfortable from the first days of his stay. This should be done in the admissions committee at the earliest stage of the applicant's acquaintance with the university. In the admissions committee besides ordinary activities to agitate children for admission, it is worth carrying out the following ones:

- joint meetings of applicants and senior students;
- meetings of applicants with the university administration;
- personal master classes in breakthrough areas of the university for applicants;
- summer scientific schools at the university during the admission committee.

So, let's talk in detail about each of the activities. Let's imagine an ordinary average applicant who graduated from school in the summer, successfully overcame all the entrance tests and is ready to enter a higher educational institution. He is concerned about all the psychological barriers that we have described above, and at the stage of the admissions committee tries to find out as much information about his university as possible. It is at this stage that the applicant should be made to understand that after admission he will not be left to himself in any case, but on the contrary, the university will accept him into its friendly and warm environment from the first day [6].

Joint meetings of applicants and senior students

The very first events during the admission committee, in our opinion, should be meetings of applicants and senior students. You will probably ask the question: "Why should meetings with undergraduates be the first important event, and not with the university administration? The administration is more competent in the field of education and will be able to answer the most outstanding and difficult questions from their experience". Psychology will help us answer this question. From the point of view of psychology, when young people move from one environment to another, it is much easier for them to perceive information from people about their own age. A similar worldview and assessment of actions play a role here, applicants differ from undergraduates much less than from the administration of a university. A completely different question is the competence of undergraduates who will talk about their Alma mater and answer questions, because we all know that the first impression is the most memorable. Therefore, when preparing for such events, the organizers should seriously and carefully approach the selection of undergraduate students.

Meetings of applicants with the administration of the university

After the applicants have learned the overwhelming amount of information from their "colleagues", the university administration at their meetings will tell about the specifics of the environment where the applicants are enrolled. It is important here not to repeat and not to touch on the topics that were discussed at the previous meeting with the students. Therefore, when preparing for such events, it is necessary to prescribe a scenario and a list of topics that need to be discussed [7].

The main criteria for evaluating these events is the number of applicants who have increased interest in the university and, according to external signs, anxiety and excitement have gone away. If after a while the meeting turned from an official to a warm friendly conversation it can be surely stated that the barriers of young people have been removed.

Personal master classes in breakthrough areas of the university for applicants

The vast majority of school graduates choose a place for higher education according to 5 main criteria:

- university rating;
- the relevance of the specialty;
- highly qualified teaching staff;
- modern level of education;
- high-tech laboratory facilities (mainly for technical universities).

If you can find out about the first 3 criteria by going to open days, talking with graduates, and reading information on digital platforms, then how can you find out about the modern level of education and high-tech laboratory facilities. To make sure that the university responds do these criteria, it's not enough just to find some information, it's necessary for future students to look at the facilities and interact with some technologies. To determine such important points as the level of education and laboratory facilities it is necessary to conduct personal master classes, where children will get acquainted with the methods of training highly qualified personnel and learn a lot of useful scientific information before starting training. These can be master classes related directly to the training specialties, psychological master classes, for example, "How to adapt to a new environment in the shortest possible time" and etc. [8].

Summer scientific schools at the university during the admission committee

Holding summer scientific schools is also one of the ways to overcome the adaptation period. Often after enrolling in a university new students have about 2 weeks to recuperate before the start of the new academic year in the status of a student of a higher educational institution. Scientific schools are an excellent tool for the development of first-year students in the educational environment and an excellent way to get acquainted with the teaching staff, as teachers in scientific schools are often full-time university teachers. And, of course, a distinctive feature of schools from master classes is the frequency of lessons. Schools can arrange the learning process in such a way that students appear every day, which allows them to adapt to a new space faster.

It is worth noting that these methods of adaptation must be carried out during the admission committee, i.e. at the very initial stage of acquaintance of future students with the future place of study. Next, we will talk about adaptation methods during training in the 1st year [9].

This method is used in many educational institutions, but is not effective in all of them—curation and mentoring. In the first year, when students join study groups, if they did not pass the adaptation methods during the admission committee, which we described above, they are usually not sufficiently immersed in the specifics of the educational institution. They do not know the material base, do not orient themselves on the territory and, most importantly, are not familiar with each other. In this case, the most effective way to unite the children among themselves and give information about the university is to assign a curator to each study group. As a rule, the curator is a senior student who is studying in a related specialty and at the same department as first-year students. He must have the following personal competencies and skills:

- good nature;
- high organization;
- sense of duty;
- communication skills;
- organizational skills;
- oratory;
- responsibility;
- openness;
- constant self-development;
- developed analytical abilities.

This is only a small part of the qualities that a curator should have. At the very initial stage, the main task of the curator is to create a warm microclimate inside the study group, teach the students how to work as a team, become real friends always ready to help each other. Secondary tasks are also very important since the educational environment is very multifaceted and the influence of various factors is also very important for the successful completion of training. Figure 3 shows the functions of the curator during the adaptation period for first-year students [10].

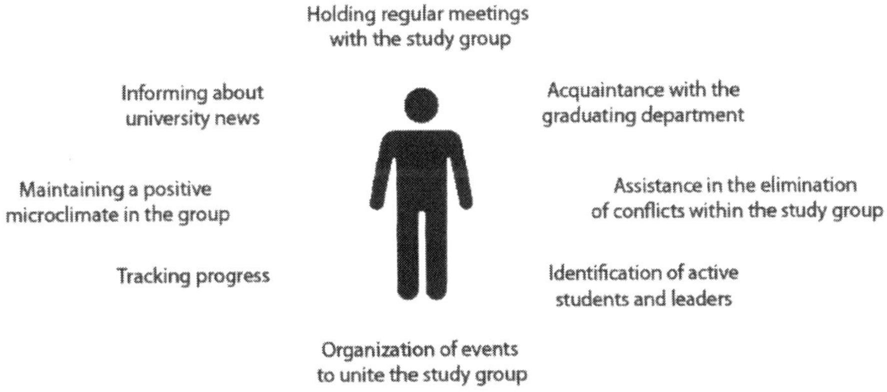

Figure 3. Functions of the curator

After completing the 1st year of study, students become more actively involved in student and scientific life. As the main part of the adaptation is completed, the curator is replaced by a mentor, whose main task is to monitor their student, as well as help him/her in personal and professional growth during training. The mentor is an employee of the scientific laboratory or department of the educational unit where the student studies. The mentor is assigned to a specific student and leads him to graduation from the university. A higher educational institution in addition to training highly qualified personnel for work in industry must form its own personnel reserve. So, the task of the mentor is to interact with his/her student and form professional qualities in him/her for further admission to the master's program and then to a PhD program to become a university teacher. Naturally, this process and the prospect of realizing oneself in higher school after graduation is not suitable for everyone, but it is one of the effective ways to grow professional staff for your own organization. Such a form of overcoming the adaptation period of first-year students has been successfully implemented at the Institute No. 1 "Aviation Technology" MAI (NRU).

In addition to curating and mentoring to overcome the adaptation period a special place is given to the educational work of students. There are separate principles of the educational process at the university:

1) freedom of choice—the right to choose must be balanced by conscious responsibility;
2) activity—transform, complement, find new connections;
3) feedback—tracking moods, level of understanding, degree of interest;
4) openness is a combination of diverse educational models and the life experience of an individual subject;
5) ideality—formation of professional motivation, demonstration of professional values based on reinforcement of personal success;
6) diagnosticity is the study of personality activity, upbringing, discipline, interests, professional orientation, sociability, commitment, organization, responsibility, needs, the level of development of the emotional and motivational sphere, self-criticism, exactingness.

Educational work at the university is divided into the following areas:

1) cultural and moral;
2) educational and professional;
3) family and household;
4) social;
5) economic;
6) physical;
7) aesthetic;
8) patriotic;
9) civil;
10) environmental.

After the end of the adaptation period for first-year students, educational work continues, as a person constantly learns and discovers something new throughout his life, and as the Soviet politician and statesman V.I. Lenin said: "Study, study and study again".

Summing up, I would like to say that at all stages of the existence of domestic and foreign education, breakthrough professions changed with regular frequency, this was due to scientific and technical progress. The world community has been mastering new technologies for years, which were discovered by great scientists and industrialists. Over the years, the methods of education have been restructured countless times in an attempt to improve the training of highly qualified personnel, but one thing has always remained unchanged—the psychological state of a person when he moves from one stage of life to another. In our case, this is the transition of a schoolchild to the status of a student. The psychological state is fundamental for its owner, but for various reasons people prefer to forget about it, focusing their attention on the personal and professional qualities of a person. In this article we have analyzed what ways there are to overcome such a psychological barrier during the adaptation of people in a new educational environment and what methods of educational work exist for the cultivation of high-quality and professional personnel.

References

[1] E.A. Gnatyshina, N.V. Uvarina, A.V. Savchenko, *Characteristics of adaptation processes at the university in the conditions of socio-cultural dynamics: Comparative analysis of adaptation processes of foreign and domestic university students*, Bulletin of the South Ural State University, **10**, 34 (2018)

[2] D.A. Prokopenko, O.S. Dolgov, B.B. Safoklov, D.B. Sukhanov, *Adaptation of a student of a higher educational institution to the educational process in the conditions of distance learning*, in Collection of the 9th International scientific and practical Internet conference "Modern challenges and actual problems of science, education and production", **1**, 476 (2020)

[3] E.L. Bolotova, A.N. Rizhov, G.M. Kojaspirova, *Pedagogy of higher education* (Prospekt, Moscow, 2021)

[4] I.A. Konstantinov, V.A. Goncharova, *Competition model in the Russian higher education system*, Education Management Review, **3(43)**, 111 (2021)

[5] E.A. Volzhenina, V.A. Kaigorodova, D.S. Shavelkin, *Methods of conducting classes in a distance learning environment*, Modern Pedagogical Education, **10**, 58 (2021)

[6] D.A. Prokopenko, V.A. Kaigorodova, *Aerospace education in Russia*, Staffing of the military-industrial complex, **11**, 164 (2021)

[7] E. Allen, J. Seaman, R. Garrett, *Blending in: The Extent and Promise of Blended Education in the United States* (Sloan Consortium, Needham, 2007)

[8] J. Mackness, M. Waite, G. Roberts, E. Lovegrove, *Learning in a Small, Task-Oriented, Connectivist MOOC: Pedagogical Issues and Implications for Higher Education*, International Review of Research in Open and Distance Learning, **14(4)**, 145 (2013)

[9] T.R. Liyanagunawardena, A.A. Adams, S.A. Williams, *MOOCs: A Systematic Study of the Published Literature 2008–2012*, International Review of Research in Open and Distance Learning, **14(3)**, 202 (2013)

[10] H. Singh, *Building Effective Blended Learning Programs*, Educ. Technol., **43**, 51 (2003)

Scientific schools of universities and character-building work

Sergey Reznik[1,*]

[1]Department SM-13, Bauman Moscow State Technical University, 5/1, 2nd Baumanskaya st., Moscow, 105005, Russia

Abstract. Scientific schools shape the image of universities, while the activity of the scientific schools enables sustainable and dynamic development of the universities. This paper is devoted to the role of scientific schools in character-building work in the context of leading aerospace engineering universities of Russia. The paper stresses the value of building the students' character based on the example of the older senior counterparts capable of transferring their experience and teaching the cutting-edge methods of educational and scientific work. Among these methods, the paper highlights a historical approach in the description of the logic and evolution of design and manufacturing solutions and the method of size effect, allowing in-depth comprehension of the true scale and composition of complex aerospace equipment. Aerospace universities of Russia implement these methods with special study laboratories called demonstration halls. Having unique specimens of spacecrafts in these laboratories helps preserve the historical memory of the forefront research of Russian scientists and engineers. Practical classes in such laboratories are a fine example of transferring the scientific legacy and positively affecting the motivation for professional growth.

1 Introduction

Character building work is an integral part of the university experience [1]. Besides regular activities within the framework of character-building work in line with the current state policy, universities play a proactive role in the character-building of the youth. The activity of these scientific schools manifested in educational programs is embodied in lecture courses and practical classes and equipping study laboratories and research centers. Also, individual student-teacher work is just as important. It can be especially fruitful when traditions are fused with new methods and means of scientific and study work [2–10].

2 The role of scientific schools in the character-building work

In the light of the ongoing discussions about new approaches to training specialists for dynamically developing industries, such as aerospace [1–6], the question of the role of scientific schools in the life of universities is relevant.

Universities have always been set apart by an intrinsic (almost invisible under the veil of routine) process of the transfer of knowledge of and experience from the elders (teachers) to

*e-mail: sreznik@bmstu.ru

the minors (students) via various forms of communication: classes, personal consultations, scientific clubs, conferences and so on. Vocational and business competencies, moral and civil stance of the universities' graduates are formed mainly within the framework of the scientific schools led by the role models of the older counterparts, the Teachers.

A scientific school is a cohesive group of one-minded researchers united by the common research topic maintaining the research continuity and having the leader well-acknowledged in the research community, as well as having significant scientific achievements, both novel and practically relevant. Thus, the quality of educational and mentoring work is directly influenced by the "health" of a scientific school and by a unique set of features setting this school apart from a regular group of teachers. Scientific schools preserve and pass over the legacy of the creative mindset, the spirit of collaborative effort, and respect for their research field and to the Motherland.

In technical universities, many scientific schools are a fine example of a caring approach to the historical memory and the transfer of the best engineering practices to the new generations of researchers and designers. The evolution of scientific schools was immortalized in literature [11–22].

For example, the scientific school of the department E-1 "Rocket engines" of the Bauman State Technical University led by professor D.A. Yagodnikov regularly hosts scientific conferences devoted to the memory of founders of two schools of research, professors V.N. Kudryavtsev and V.N. Polyaev. In addition, the department E-1 re-published tried and tested, yet still relevant, study books, for example, the study book on liquid rocket engines written by M.V. Dobrovolsky.

The department SM-1 "Spacecrafts and launch vehicles" of the Bauman Moscow State Technical University is one of the first departments making engineers specializing in the design and manufacture of guided long-range ballistic missiles. The SM-1 department has a scientific school of dynamics and strength of thin-walled and spatial structures led by Professor V.N. Zimin. The founders of this scientific school, V.I. Feodosiev, K.S. Kolesnikov, N.A. Alfutov, L.I. Balabuh, S.A. Alekseev and V.I. Usyukin had made a significant contribution to developing methods for designing and testing shell and truss structures. The SM-1 department regularly hosts conferences and workshops in memory of famous teachers and scientists and published many books preserving the memory of the pioneers of rocketry and cosmonautics [19–22].

Table 1. Comparison of heat loads in modern aviation, rocketry, and space industry

Parameters/Loads	Aviation	Rockets	Spacecrafts
Flight density, m/s	Up to 1000	7000	7900–60000
Heat flux density, kW/m^2	Up to 2	$5 \cdot 10^3$	$< 2 \cdot 10^4$
Pressure, MPa	< 0.7	< 20	< 10
Duration of the atmospheric stage, hours	< 24 and over that value with re-fueling	0.01	0.5

In Bauman Moscow State Technical University, a scientific school of composite materials' science had started taking shape in the mid-1950s influenced by the roaring development of rocketry and cosmonautics. The unique difference affecting the choice of design and manufacturing solutions for aviation and rocket-space structures was based on a significant difference in in-flight heat and force loads acting on planes, missiles, and spacecrafts (table 1).

It was only the use of composite materials for thermal protection that helped overcome so-called "thermal barriers" that haunted designers of the guided long-range ballistic missiles, crewed spacecrafts, and automatic interplanetary stations. The adoption of composite technologies has paved the way towards solving the tough problem of creating mixed solid-fuel missiles with composite bodies and transport and launch containers. Without new carbon-based composites, there would be no reusable "Buran" orbiter, light and precise reflectors of onboard space antennas, and the load-bearing meshed space structures having the top weight-to-strength ratio.

In the framework of the study course "The history of the scientific school of composite materials", the students of the department SM-13 "Rocket and space composite structures" get to know the names of the authors of pioneering solutions in the area of design, manufacture, and testing of composite structures of aircraft and spacecrafts. Many of these authors are Bauman Moscow State Technical University graduates. Among them are the chief designers of the industry's leading enterprises: S.P. Korolyov (OKB-1), P.A. Tyurin (TsKB-7—KB "Arsenal"), B.N. Lagutin (NII-1—MIT), V.D. Protasov (KBM—TsNIISM), designers K. P. Feoktistov (OKB-1), A.B. Mitkevich (KBM—TsNIISM), V.V. Vengerskii (NII-125—LNPO "Soyuz"), material scientists G.G. Konradi and V.N. Iordanskii (NII-88) and many others. Also, a significant contribution in new research and study courses was made by professors E.A. Satel, V.I. Feodosiev, N.A. Alfutov, I.M. Bulanov, A.K. Dobrovolsky, E.F. Kharchenko, M.A. Komkov, V.M. Kuznetsov, G.E. Nekhoroshikh, V.A. Shishatsky, A.A. Smerdov, G.B. Sinyarev, V.A. Tarasov, V.I. Usyukin, V.S. Zarubin, P.A. Zinoviev and others.

In Russia, the creation of technically complex composite structures, perfect both in terms of weight and functionality, demonstrates a high level of Russian science and engineering serves as a motivational example and boosts the students' confidence that they chose the right "vocational trajectory".

In acquiring professional competencies, practical classes solidifying the student's memory of pioneering engineering solutions for complicated structures are essential.

3 The size effect of studying real specimens of equipment

The classes devoted to studying the hardware of rockets and spacecrafts in special laboratories (demonstration halls) are some of the effective forms of professional training. It should be noted that these classes are based on the organic synthesis of the historical approach and the size effect in training.

The demonstration hall of the department SM-1 of the Bauman Moscow State Technical University has been operational for more than 70 years. It was equipped under the supervision of the associate member of the USSR Academy of Sciences V.I. Feodosiev assisted by the industry and the department's teachers (B.K. Kovalev, V.I. Vorotnikov, I.P. Medov, N.I. Popkov, N.A. Suratov, and others). As early as in the mid-1950s, the hall had been equipped with a then-novel rocket R-2 (8Zh38). Over time, the hall had become too small for new equipment specimens, so the demonstration hall was moved to a new building in the suburban base (currently the Dmitrov branch of the BMSTU) in the settlement Orevo of the Dmitrov district of the Moscow Region (figure 1).

Within the framework of a legacy-based approach, the students can face the journey of creative discoveries of the primary designers of long-range ballistic missiles, S.P. Korolyov, V.N. Chelomei, and M.K. Yangel. Evidently, reading books, studying interactive materials, and even going on excursions is nowhere near this hands-on experience of the study materials relevant to the engineering specialty. The possibility of having a close look on a real rocket fuel tank, touching the rocket fixtures and thermal protection coatings gives one an insight into the designers' logic, then-innovative material science, and manufacturing innovations and helps form one's judgment of how proper those designs were.

Figure 1. Laboratory of the department SM-1 "Spacecrafts and carrier rockets" in the Dmitrov branch of the Bauman Moscow State Technical University

Figure 2. Laboratory of the department 601 "Space systems and rocketry" of the Moscow Aviation Institute

The Moscow Aviation Institute (national technical university) department 601 "Space systems and rocketry" founded in 1959 with the participation of the academician V.P. Mishin boasts a rich experience and wide capabilities for training specialists. alleviate the formation of vocational competencies.

This department, led by the academician O.M. Alifanon for more than 30 years, has a demonstration hall and study classes with unique specimens of Russian space equipment, including the lunar spacecrafts 7K-LK, the "Soyuz" type descent modules (figure 2). Due to the size effect, the laboratory practice and the study of real equipment specimens actively.

Figure 3. Laboratory of the department A-1 "Rocket building" of Baltic State Technical University "Voenmeh" n.a. D.F. Ustinov

Figure 4. A ballistic missile Polaris A3 in the Royal Air Force Museum Cosford, UK

This year, the department A-1 "Rocket Building" of the Baltic State Technical University "Voenmeh" n.a. D.F. Ustinov led by professor V.A. Borodavkin celebrated its 75th anniversary. The demonstration hall of the department is in an exemplary state, and it is also used for implementing the historical approach and the size effect in training (figure 3).

There is a long list of universities with an excellent space equipment collection. However, Samara National Research University named after S.P. Korolyov, South Ural State University, Peter the Great Military Academy of the Strategic Missile Forces, A.F. Mozhaysky's Military-Space Academy deserve special attention for their collections.

It should be noted that Western universities, as a rule, use state or enterprise museums for that purpose, for example, the Boeing Museum in Seattle, the Royal Air Force Museum in London and in Cosford, the Museum of Science and Industry in Manchester having unique specimens of planes, rockets and jet engines, the ESA exposition in the European Space & Technology Centre (ESTEC), Noordwijk and so on figure 4.

Museums of equipment are a necessary thing. However, one should bear in mind the main rule of a museum: do not touch the items! In this case, the size effect is quite different.

4 Conclusion

The prospects for the development of rocket science and cosmonautics (new reusable manned spacecraft, small and ultra-small spacecraft, the construction of orbital and planetary structures, space tourism, etc.) serve as an incentive to improve the forms and methods of teaching students of technical universities. Scientific schools of aerospace universities in Russia have a powerful potential in educational work. It is advisable to preserve time-tested effective forms of education that combine a historical approach and a size method. Reasonable conservatism along with the development of advanced scientific methods and tools will serve as a guarantee of high quality training of a new generation of specialists.

References

[1] D.A. Kozorez (eds.), *Aerospace education in Russia. Staffing of the military-industrial complex* (MAI Publishing House, Moscow, 2021)

[2] V. Mayorova, *Integration of educational and scientific-technological areas during the process of education of aerospace engineers*, Acta Astronautica, **69(7–8)**, 737 (2011), DOI: 10.1016/j.actaastro.2011.04.007

[3] N.I. Sidnyaev, *Contemporary day discussion on the concept of elite engineering education*, Engineering Education, **17**, 14 (2015)

[4] J. Rodriguez, A. Laverón-Simavilla, J.M. Del Cura et al., *Project based learning experiences in the space engineering education at Technical University of Madrid*, Adv. Space Res., **56(7)**, 1319 (2015), DOI: 10.1016/j.asr.2015.07.003

[5] A.Yu. Lutsenko, I.R. Shafikova, A.L. Galinovskiy, *Training features of the engineering personnel for the high-tech sector of the economy in the UK*, Handbook. An Engineering J., **10(247)**, 50 (2017), DOI: 10.14489/hb.2017.10

[6] G. Shcheglov, V.I. Mayorova, V. Zelentsov, *Integration of controlled joint mixed-year teamwork for designing ultra-small satellites into the specialist degree educational program*, Int. J. Mech. Eng. Technol. (IJMET), **10(7)**, 82 (2019)

[7] V.I. Mayorova, O.K. Bannova et al., *Importance of joint efforts for balanced process of designing and education*, Acta Astronautica, **111**, 249 (2015)

[8] O. Bannova, V. Mayorova, *Building trans-national and multi-disciplinary academic curricula through adaptation of a project-based approach*, AIP Conf. Proc., **2171**, 140002 (2019)

[9] L.V. Juravleva, V.A. Shakhnov, A.I. Vlasov, *Adaptation of professional engineering training to the challenges of modern digital production*, Advances in Intelligent Systems and Computing, **1134**, 623 (2020), DOI: 10.1007/978-3-030-40274-7_59

[10] I.R. Shafikova, *Reading in English at an engineering university. A matter of importance*, AIP Conf. Proc., **2318**, 020015 (2021), DOI: 10.1063/5.0035983

[11] L.I. Volchkevich, I.L. Volchkevich, *Bauman Moscow State Technical University: History and modernity* (Bauman MSTU Publishing House, Moscow, 2005)

[12] I.B. Fedorov, K.S. Kolesnikov, *Scientific schools of Bauman Moscow State Technical University. History of development* (Bauman MSTU Publishing House, Moscow, 1995)

[13] N.P. Alyoshin, G.P. Pavlikhin, I.B. Fedorov, *Academician Georgy Aleksandrovich Nikolaev. Pages of the life of an outstanding scientist, teacher and organizer of a higher school* (Bauman MSTU Publishing House, Moscow, 2002)

[14] G.N. Antsupova, G.P. Pavlikhin, *Rectors of Bauman MSTU (1830–2001)*, 2nd ed. (Bauman MSTU Publishing House, Moscow, 2002)

[15] I.V. Barmin, O.P. Matveeva, V.V. Chugunkov, *Academician V.P. Barmin and the Department of "Launch rocket systems": To the centenary of the birth of V.P. Barmin and the 50-th anniversary of the Department* (Bauman MSTU Publishing House, Moscow, 2009)

[16] V.D. Protasov, *To the 90-th anniversary of his birth* (RPM Publishing House, 2021)

[17] V.N. Gushchin, E.L. Kalyazin, Yu.A. Matveev, V.K. Serdyuk, *Department of Space Systems and Rocket Engineering of the Moscow Aviation Institute (state technical university). To the 45-th anniversary of the foundation*, ed. by O.M. Alifanov, V.N. Yarotsky, D.B. Payson (MAI Publishing House, Moscow, 2004)

[18] V.I. Ruslanov, P.P. Afanasyev, *Moscow Aviation Institute. Facts, figures, documents. Reference materials* (MAI Publishing House, Moscow, 2009)

[19] P.P. Afanasyev, V.I. Ruslanov, I.K. Turkin, *Sixty years of training rocket engineers at MAI. To the 60-th anniversary of the department 602. Department of "Aviation and Rocket Systems" (1952–2012)*, ed. by I.K. Turkin (MAI Publishing House, Moscow, 2012)

[20] V.I. Usyukin, A.I. Ganichev, S.V. Reznik (eds.) *M-1: 50 years after the start* (Bauman MSTU Publishing House, Moscow, 1998)

[21] V.N. Zimin, V.N. Eliseev, S.V. Reznik et al., *SM-1: The flight is normal. To the 100-th anniversary of the birth of V.I. Feodosiev*, ed. by V.N. Zimin (Strannik, Moscow, 2016)

[22] Kovalev B.K., *I thank my fate. Letters to a friend* (Publishing House "RTSoft"—"Kosmoscop", 2016)

Axiological approach in the university's transformation based on the use of digital and remote educational technologies in pandemic and normal conditions

Tatyana Sazonova[1,*], *Olga* Khristodulo[2], *Aida* Kiryakova[3], and *Nazira* Dusheyeva[4]

[1]Kumertau branch of Orenburg State University, 3B, 2nd lane Sovetsky, Kumertau, 453300, Russia
[2]Ufa State Aviation Technical University, 12, K. Marx st., Ufa, 450008, Russia
[3]Orenburg State University, 13, Pobedy prospect, Orenburg, 460018, Russia
[4]Kyrgyz Academy of Education, 25, Erkindik blvd., Bishkek, 720010, Kyrgyzstan

Abstract. The digitalization of education due to constant transformation is considered as an existing reality and at the same time is a challenge of a modern age. It will invariably lead to changes in the content of education, methods and organizational forms of educational work. The article considers the issues of the university's transformation, describes the experience of using digital technologies during the mass transition to the remote studying on the example of two regional universities. Issues of axiology in transformation of education are considered.

Computers have been established in our lives quite long and substantially, although initially they were created as a means of computing but now it is impossible to imagine an educational process of schools and universities without digital technologies [1]. In two decades the computer has turned from a powerful bulky to a portable device that can perform a huge number of functions and combine various previously incompatible functions. It is a means of communication, information transfer, remote access and control systems, and even health control etc. In fact this list may be very large, many operations and inventions are impossible without a computer. In the modern world, the best personal computers are involved in many areas of production. They help in education and science, serve in the country's defense system and represent an integral part of medicine [2].

The introduction of networks made it possible to turn a computer from a computing device into a communication device. The next century in the evolution of humanity began with the emergence of the Internet. It is the age of "information technology", the age of computers and computer networks development; it is the time when the information delivered on time represents a high value. Today we have learned to communicate with a computer in a language that it understands. This device in any difficult situation is ready to suggest ways of activity in a specific case. Thus speaking of digital transformation in education, we can confidently say that learning is a way of human intellectual development, including the usage of artificial computer intelligence [3].

Today, like 100 years ago, the knowledge production depends on the nation's intelligence and determines its fate. According to the strategy of the Digital Transformation of the Science and Higher Education Sector, developed as part of the implementation of the President of the Russian Federation Decree of July 21, 2020 No. 474 "On National Development Goals of the Russian Federation for the Period until 2030", the achievement of "digital maturity"

* e-mail: SazonowaTV@kfosu.edu.ru

of the economy's key sectors and social sphere is indicated as one of the determinants of the national goal "digital transformation". It involves updating the planned educational results, the content of education, methods and organizational forms of educational work, and evaluating the achieved results in a rapidly developing digital environment for dramatically improvement of each student's educational results (individualization of training) [4].

Transformational processes in education have begun around the world. They are also needed in national education. The digital economy requires that each trainee (and not just the best) master the competencies of the 21st century (critical thinking, the ability to self-study, the ability to use digital tools, sources and services in their daily work) and can creatively (not according to a template) apply existing knowledge in a rapidly developing digital environment [5–9]. The new Federal State Educational Standard has already set the task of development of each student's ability to manage his own learning. The digital transformation of education must solve this problem.

Moreover, the most powerful breakthrough in this sphere was made in 2020, which became a challenge for everyone. The pandemic has transformed everything including politics, economics, medicine, science, education, and that is the most important it has changed worldview.

March, 2020 was a serious test for the educational system. It was necessary to restructure the educational process from the usual classical verbal communication "Teacher–student" to the remote format "Teacher–computer" assuming students as a target audience. In the transferring of all forms of the educational process to the online environment from lectures to all types of practical classes, it has been found that the educational process is not ready for such a fundamental restructuring. First, there is the lack of the actual level of academic subjects' development in the form of electronic educational resources and the insufficient level of teacher's professional training for the full and high-quality use of digital platforms and services in the educational process.

In this situation quality and content of the educational process is sharply reduced because it is not possible to use information and communication technologies in fully scale, but the volume of information itself is steadily and rapidly growing, creating an unfulfilled process in the loss of the received knowledge quality [10].

An analysis of the general and qualitative educational achievement of full-time students was carried out on the example of two regional universities—Federal State Budgetary Educational Institution of Higher Education "Ufa State Aviation Technical University" (USATU) and Federal State Budgetary Educational Institution of Higher Education "Orenburg State University" (Kumertau branch of OSU). The results of the analysis are represented in figure 1.

As it can be seen from the diagrams presented in figures 1 and 2 the qualitative and general academic achievement in two universities on different training programs and related to various areas of activity during the mass transition to distance learning during the 2020 pandemic, has significantly decreased.

The results of the session (see figures 1 and 2) showed that new pedagogical technologies are needed to create individual educational trajectories, evaluate the professional situation based on innovations and basic principles of labor functions, motivate in-depth critical study of professional experience [10, 11]. A large concentration of symbols, meanings, communications, actions, results, and, therefore, another education is needed.

It is explained by the following disadvantages in distance education:

1) mandatory availability of technical equipment that meets the requirements of conducting the educational process in a remote format;
2) availability of electronic lecture courses and measurement materials for intermediate and final academic efficiency evaluation;
3) lack of laboratory training using a specialized equipment;

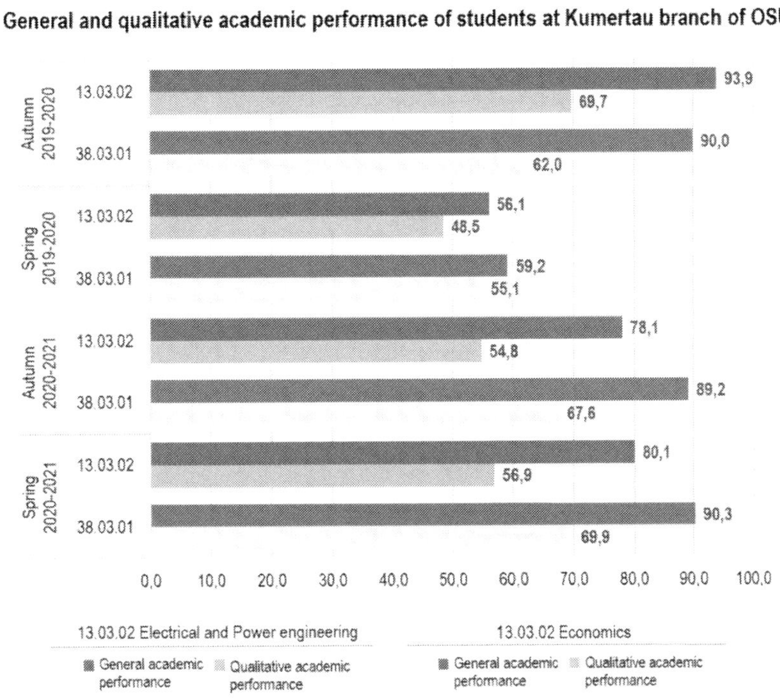

Figure 1. The analysis of the general and qualitative academic achievement of full-time students in regional university—Kumertau branch of OSU

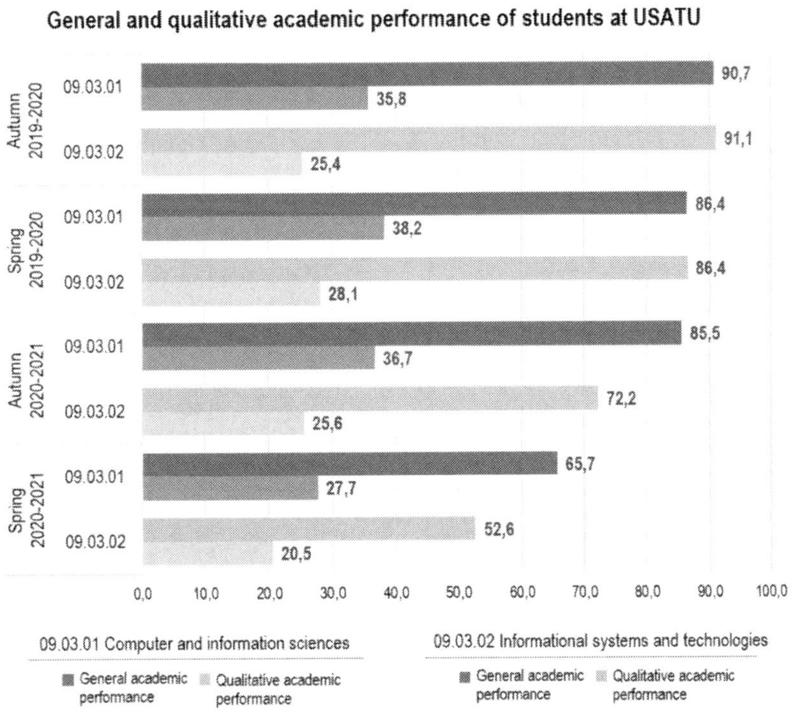

Figure 2. The analysis of the general and qualitative academic achievement of full-time students in regional university—USATU

4) the difficulty of supporting student motivation and interactivity during classes;
5) lack of direct communication with students;
6) establishing the user's identity during knowledge testing;
7) the dependence of learning success on technical skills.

Turning to global trends in education, we are clearly aware that education plays the role of the genetic matrix of culture, being an essential element of the mechanism for translating the information code in the structure of culture, ensuring its reproduction and development. Axiology (science of values) is in general line with the search for answers to acute questions and challenges of our time [12].

Axiology considers objective reality combining the laws of the objective and subjective relationships, current and potential, external and internal, essential and proper, traditional and innovative, natural and random, stable and dynamic in the capacious axiological dimension of the modern world. As the axiological dominant of the modern specialist training, it is necessary to designate computerization and its strongest impact on the student's modern world [13]. Without modern attributes of communication, teaching, leisure activities such as gadgets, a modern student is unthinkable, he feels flawed, confused, suspended. Education cannot ignore this fact; therefore, it is necessary to look for options for using this phenomenon as an instrument of strategy and tactics of interaction between a teacher and a student in the university's educational process.

The axiological approach involves considering not only the external factors of the university's transformation, not only external circumstances and situations of the modern person's life, but also the internal peculiarity of the student's personality, his needs, motives, value orientations, aspirations, expectations and claims.

COVID-19 was a major shock to the educational system around the world. Russian education was no exception, but there are also positive points in this situation: people began to master new professions, educational structures, learned to work with new formats, and the model of "mixed learning" from the perspective becomes a reality [8].

The pandemic changed the Internet. In 2020 during the period of self-isolation many professions and activities went online and will not return completely offline. According to the results of the Institute of Statistical Research and Knowledge Economics of the Higher School of Economics (HSE University) about three quarters (76%) of respondents began to use digital tools more often to solve various everyday problems. Half of respondents (49%) installed additional applications and programs. Every third (34%) has mastered a new skill, and 48% plan it in the near future [14, 15].

To meet the needs for mass digitalization providers have significantly increased the speed, quality and Internet coverage over the past year. At the beginning of 2020 more than 4.5 billion people used the Internet, that is, almost 60% of the world population is already online and there is a minimization of digital inequality. At the beginning of the transition to distance learning throughout the country (March, 2020) there were problems with both Internet access and the availability of telecommunications, then in 2021, according to the Ministry of Digital Development of the Russian Federation, mobile communications will appear in 1198 villages, and settlements inhabited by 100–500 people. Mobile communication of the 4G/LTE standard will become available to residents of these settlements; it will also provide sustainable access to the Internet.

There is a significant increase in online education, and the concepts of "education" and "learning" are eroded and embedded in youth leisure through the Internet. New skills for a modern student become prevalent and essential and a new quality of education such as "Connectivity" is formed. That is the ability to be connected. A habit of information concentration is created. It is necessary to receive several professions and study continuously since the development of technology and the reduction of the professions' life cycle have led to the obsolescence of the classical school-university education model. It was replaced by the concept of lifelong learning [16–18].

Some digital practices have become more popular in conditions of forced self-isolation, but have also shown the possibility of replacing traditional methods of activity. The active use of digital services and devices has stimulated people to improve their skills, which may lead to an increase in digital literacy rates in the near future.

The competence "ability to work in a team" and in an online format, is increasingly appreciated by the employers.

There is clear competition in the field of educational services in the rapidly developing conditions of digitalization. So the share of the population of the Russian Federation according to the Rosstat (Federal State Statistics Service `https://www.gks.ru/`) at the end of 2020 the population of Russia was 145,963,350 people. In 2020 the population of Russia increased by about 59,820 people. Taking into account that the population of Russia at the beginning of the year was estimated at 145,903,530 people, the annual growth was 0.04%, and the natural population growth was 167,789 people. As a result, you should not expect an increase in the number of students and schoolchildren. Sometimes it is enough to take short-term online courses to master new professions and soon most of humanity will study remotely, since now there are all the technical possibilities for creating online courses that are not inferior to classical (including university).

Considering the current situation and the primary importance of improving the quality of bachelor's professional training for the requirements of the economic sectors the main task arises. This task is to improve the individuals development, both trainees and teachers in the field of the digitized environment through the development and implementation of new technological, methodological and information opportunities with the transition to the formation of a unified information-and-communication educational environment.

Digital reality education is a new research opportunity, high students' educational performance and innovations in learning and research.

Other kinds of technologies, other types of people, and another culture are needed to achieve these goals. Digital transformation is a series of profound and coordinated changes in culture, staff and technology that engender new educational and operational models and lead to the transformation of institutional operations, strategic directions and value propositions [6, 19].

The peculiarities of the student's value self-determination in the axiosphere of the Internet are relevance, speed, intensity, multi-vector, great freedom, a large number of opportunities and innovative technologies, a specific space of self-determination. The Internet can be seen as a very rich space and a means of developing creative potential only when the person is focused on humanistic values, the values of self-realization, self-improvement, and ready to create new, unique and beautiful things.

Determining the coordinate system in which the main meanings of the educational transformation of the university are presented and revealed axiology focuses on identifying priorities, dominants as basic vectors of the socio-cultural practices development in the digital world [6, 12]. The intensity and speed of the changes taking place in the context of computerization concerns every person. Intense searches for solutions to the optimal combination of the traditional and innovative nature of the educational activities of the university are carried out. The educational community has to estimate the obvious difficulties, risks, identified problems, see unforeseen, uncertain trends and circumstances. It will be necessary to re-understand the traditional methods and fundamental foundations of the education development in order to be ready to realize the set goals of educating and teaching students in a new, radically changed situation of educational development.

In this regard we outline the prospects for the educational digitalization (figure 3).

1. Digital transformation is a system, a process, a series of objective, profound and co-ordinated changes in culture, personnel and technologies that involve new educational and operational models and lead to the transformation of strategic directions and technological solutions. This will certainly determine positive developments in education.

2. Environment. The saturation of university practice with new information resources and technologies is unopposed (despite the difficulties that are inevitable). This means that it is necessary to continue to create a special educational environment that the best educational organizations have already been able to provide for teachers and students. It requires broad-band Internet access to work with big data for productive educational interaction between teachers and students on Zoom, Teams, etc.

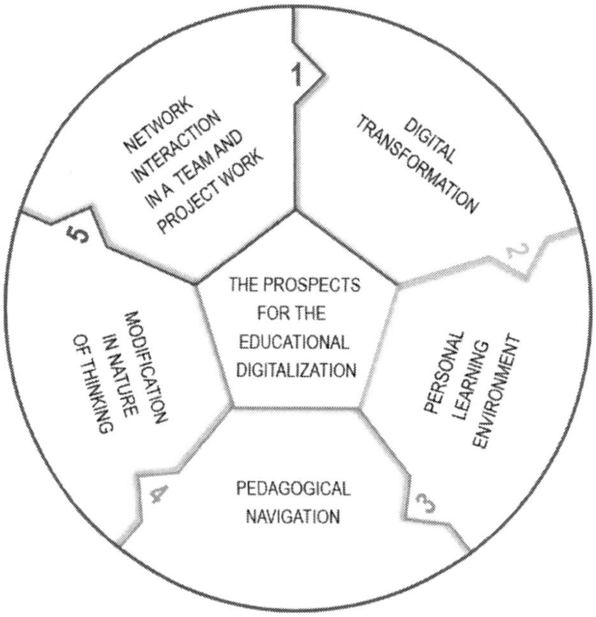

Figure 3. Model of digitalization of education

The essence of the digital transformation of education is that there is a great opportunity for everyone to achieve the necessary educational results through the personalization of the educational process, including the use of artificial intelligence methods and virtual reality tools. The usual tasks of the IT educational environment are supplemented by new ones. For example, the problem of creating a teacher's personal environment as an axiological resource for improving the quality of education becomes relevant. Personalization of education today is a necessary attribute of the educational system quality. Competence in the field of personal knowledge management today cannot be dispensed with, since the teacher needs to create his own Personal Learning Environment (PLE).

The personal educational environment as a learning environment is focused on the subject of education. A personal educational environment is a set of the educational process objects (content, forms, methods, means of training and educational communications) obtained from the information and communication educational environment by adaptation in accordance with the goals, content and planned results of training, the needs and abilities of the student and acting as a means of individual personalization (S.K. Vasilchenko). The teacher's personal educational environment is a resource of training and interaction with the students' personal learning environments and the colleagues' personal environments through which the teacher implements the achievement of educational goals and professional activities.

3. Content for learning knowledge requires pedagogical navigation. These are the basic guidelines for students to design their own knowledge. It is stated that information turns into real knowledge only with the hard thinking, the application of theoretical knowledge in practice, and in the creation of various projects.

Knowledge implies understanding. In this regard, difficult questions arise in the educational process: What is to be taught in the university auditorium and what will the student learn on his own? Will everyone master the educational material himself or herself or does he/she need the constant assistance of a teacher? The content of education today changes not only quantitatively reducing the volume of reported information, but changes functionally. "Why do I need this?" the student constantly asks himself. He cannot fully assess the feasibility and prospects of the educational process. Much seems superfluous to him. The attitude to the value of knowledge, the motivation of knowledge becomes decisive in the success and quality of education.

The rapid development of the Internet, the rapid spread of social networks like Facebook, Instagram and Twitter, the profound impact of these phenomena on the social, cultural and commercial activities of people, as well as the Internet and social media significantly change the learning process. It made teachers review approaches and methods, finding ways to work with new technologies to stimulate learning regardless of the academic subjects. The driving forces for the Internet development were not only innovation and technology, but also the release of the creative fantasy of millions of users who are quenching their growing needs, giving rise to further advanced technologies [8, p. 58].

4. The nature of thinking has already changed significantly. The nature of the formation of neural connections in representatives of generations Y, Z, AlFA is already different.

On the one hand is the frame perception of information, the dominance of visualization, clip thinking, the speed of the switching processes from object to object, irritation in the absence of quick answers to interesting problems (the teacher does not have time (as quickly as Google) to satisfy lightning-fast emerging and rapidly fading cognitive needs, and interests).

The emphasis in teaching the digital transformation of education shifts to the development of new abilities: abilities for analysis, expertise and transfer of learned knowledge and skills to new situations. Many digital technologies have a rich didactic and axiological (significant for education) potential, the characteristic of which is the freedom to search for information in a global information network. The goal of learning is to form and develop the students' ability to learn, understand the logic of new solutions that move science forward, form the educational independence of schoolchildren and students, and develop their abilities and personal potential.

5. Network. Information and communication technologies in the educational process radically change the nature of educational activities. The horizons of applicability of the acquired knowledge expand significantly with a new understanding of the of project and teamwork capabilities in network interaction. The network optimizes critical thinking, reduces the time of cooperation, and connects the "developed blocks" of the project into a whole by the Lego type. It saturates the process of communication emotionally. It provides opportunities to attract remote project participants and experts to create and evaluate products of creative joint activities. However, the role of the teacher as a moderator of activities and the project manager increases. The success or failure of this type of activity depends not only on the professional competence of the teacher, but also on his axiological literacy (the ability to create an atmosphere of passion, interest, corporatism). This problem stems from a relatively new phenomenon that is the joint creativity of students in the network. In recent years the collective design work of students in the Network using a number of services such

as Google Docs, Zoho Writer, Teamer and a number of others in the school educational process is used. These services are originally designed for collective creative activity and allow simultaneous work with text, graphic and multimedia material by many users involved in the collaborative project [2].

Many digital technologies have significant didactic potential with the following characteristics:

- Freedom of the information searching in the global network.
- Personality. It is the presence of unlimited opportunities for personal adjustment for the needs and characteristics of each student, including the choice of the material presentation method, the level of complexity, the pace of work, the number of fixing repetitions, the nature of educational assistance, partners, game entourage.
- Interactivity. It is the ability to provide multi-entity in the process of communication and interaction.
- Multimedia. It is the ability to use comprehensively various channels of perception (auditory, visual, and motor) in the educational process.
- Hypertext. It is the freedom of movement through the text, compressed presentation of information in the form of infographics, the text modularity and its optional continuous reading, the reference nature of the information, the folding-deployment of information, the use of cross-references.
- Subculturalism. It is the conformity with the usual image of self-development.

With a well-organized work of a student team united by a common task the speed of work (creating fragments, correcting them, making corrections, creating several solutions) can be rather high. Therefore, the teacher who implements such a project with his students and observing the dynamics of joint work perceives it as the activities of not individual students but a certain "collective" subject.

The key unit of the modern age is no longer a separate qualified employee as "a person in his place" but a team capable to solve tasks of a design or functional nature effectively. The processes of "compressing time and space" caused by the spread of digital telecommunications, globalization, the emergence and spread of new models of labor organization (coworking, remote offices, distributed project teams, freelance, crowdsourcing, etc.), and the professions' convergence present fundamentally different requirements for workers, including those related to the willingness to work in conditions of uncertainty.

The above-mentioned technological factors and personal changes in the conditions of digitalization set to education a new vector for the development of pedagogical methods and technologies. It inevitably will lead to educational transformation.

References

[1] *Shaping the Fourth Industrial Revolution*, Siemens AG [Electronic resource], URL `https://rn.siemens.com/en/innovation/research-topics/shaping-the-fourth-industrial-revolution.htm` (citation date: 2021-07-27)

[2] J. Manyika, M. Chui, M. Miremadi, J. Bughin, K. George, P. Willmott, M. Dewhurst, *A future that works: automation, employment, and productivity* (McKinsey Global Institute, 2017)

[3] A.Yu. Uvarov, S. Wang, C. Kang, H. Su, P. Cao, S. Jiang, Yu. Zhan, S. Zhu, *Problems and prospects of digital transformation in Russia and China*, in Materials of the II Russian-Chinese conference "Digital transformational image and artificial intelligence", 26–27 September 2019, Moscow, Russia, 155 (2019)

[4] *Forecast of socio-economic development of the Russian Federation for the period up to 2024 (approved on Session of the Government of the Russian Federation dated 19.09.2019)*, Ministry of Economic Development of the Russian Federation [Electronic resource], URL https://www.economy.gov.ru/material/file/450ce3f2da1ecf8a6ec8-f4e9fd0cbdd3/Prognoz2024.pdf (citation date: 2021-07-27)

[5] A. Vitkovsky, *Transformation of the education system: why and how it happens*, Directory-online [Electronic resource], URL https://medium.com/direktoria-online/brovkina-9c7cf1e2f423 (citation date: 2021-04-05)

[6] T.V. Sazonova, *Transformation of education as a phenomenon of objective reality* (Publishing House "University", Orenburg, 2021)

[7] A.V. Kiryakova, N.A. Kargapoltseva, I.D. Belonovskaya, *Scientific and Pedagogical Projections of the Transformation of University Education*, Higher Education in Russia, **29(8–9)**, 155 (2020)

[8] M.D. Schelkunov, *Universities of the new generation*, Bulletin of Economics, Law and Sociology, **1**, 187 (2017)

[9] M. Barber, K. Donnelly, S. Rizvi, *An Avalanche Is Coming. Higher Education and the Revolution Ahead*, Educational Studies Moscow, **3**, 152 (2013)

[10] A.V. Kiryakova. E.A. Garaeva, *Development of information and communication skills of university lecturer in conditions of digitalization of educational environment*, Vestnik of the Orenburg State University, **2(230)**, 30 (2021)

[11] A.V. Kiryakova, V.V. Moroz, *Digital educational environment—axiological student creativity driver*, in University complex as a regional center for education, science and culture: Materials of the All-Russian Scientific and Methodological Conference (with international participation), 25–27 January 2021, Orenburg, Russia, 3417 (2021)

[12] A.V. Kiryakova, V.V. Moroz, *Axiological Dominants of Personality Development in the Digital World*, in Priorities and values of individual education and development in modern society: Materials of the All-Russian Scientific and Practical Conference, 29 October 2020, Ryazan, Russia, 73 (2021)

[13] V.V. Polyakova, K.S. Fursov, *Digital practices of Russians during the period of self-isolation*, Institute of Statistical Research and Knowledge Economics—National Research University "Higher School of Economics" [Electronic resource], URL https://issek.hse.ru/news/438496284.html (citation date: 2021-07-27)

[14] G.V. Osipova, *Competency approach in education*, Multifunctional system "Informio" [Electronic resource], URL http://www.informio.ru/publications/id1961/Kompetentnostnyi-podhod-v-obrazovanii (citation date: 2021-07-27)

[15] D.P. Zavodchikov, E.F. Zera, D.P. Zavodchikova, *Innovative professional space of people* (Publishing House of Russian State Professional Pedagogical University, Yekaterinburg, 2014)

[16] S.G. Kashina et al., *Permanent nature of transformations in modern university educational environment: didactic aspect of transgression*, EurAsian J. Biosci., **14** (2020)

[17] S.V. Tetersky, *Pedagogical accompaniment of clip accompaniment to a modern youth*, Values and meanings, **1(29)**, 56 (2014)

[18] O.F. Piralova, V.M. Filippov, E.A. Beloglazova, *The reverse component of the formation and development engineering universities teachers' professional competence*, Higher education today, **4**, 27 (2017)

[19] A.A. Akhayan, E.V. Piskunova (eds.), in Collection of articles of the I International Scientific and Practical Conference "Educational dynamics of network personality", 30 January 2018, St. Petersburg, Russia (2018)

Professional precedent texts in teaching English for aerospace engineers

Inna Shafikova[1,*]

[1]Bauman Moscow State Technical University, 5/1, 2nd Baumanskaya st., Moscow, 105005, Russia

Abstract. The paper discloses the importance of professional precedent texts for teaching English to future engineers and researchers at a Master's degree program at a technical university. Three types of precedent texts found in scientific and technical articles are identified: classic precedent texts, term-type precedent texts, and professional precedent texts. It is stated that professional precedent texts have the greatest teaching potential. The most common genres of precedent texts for aerospace graduate students are review articles and original research articles in peer-reviewed journals. Criteria for the selection of professional precedent texts essential for teaching profession-oriented reading in aerospace departments of engineering universities are provided.

1 Introduction

Modern requirements for the aerospace industry development in the Russian Federation call for professionals of a new formation, capable of becoming a significant part of the global research and industry community. In order to function in the international engineering market, specialists must freely navigate the flow of scientific and professional information, which is impossible without mastering English as a second language. Professional precedent texts (PPT) are an important component of a specialist's linguistic persona. In order to answer the question of whether all texts of scientific and engineering discourse can be considered as precedent, what are the features that distinguish professional precedent texts from a huge array of scientific and engineering texts, let us turn to the history of the concepts of precedency and precedent texts.

Yu.N. Karaulov defined precedent texts (PT) as those that have emotional and cognitive significance for the user, classic or "textbook" texts, widely known and constantly present in discourse [1]. Initially, only quotations, the names of characters, the names of written works' and their authors' names were considered as the precedent texts. Subsequently, the boundaries of the precedent text category were expanded by including the texts that were significant for a narrow circle of people or known for a relatively short time [2]. In addition, the levels of precedence were identified depending on the scope of people using them. These levels are as follows: an individual, a member of society, a member of a national-cultural community and a member of the human race [3]. It is the existence of precedence at the level of an individual and at the level of a member of a certain community that makes it possible to consider non-classic, specialized texts, including scientific and professional texts

*e-mail: shafikova_iina@mail.ru

as precedent texts. Initially, and this was separately noted by Yu.N. Karaulov, scientific and professional texts were excluded from precedent texts due to the narrow specialization and low value of the former for a large part of society.

2 Types of professional precedent texts

Analysis of research on precedent phenomena in scientific and engineering discourse allowed us to identify three main approaches to the concept of a professional precedent text. It should be noted that in our study we focused on scientific and engineering precedent texts, without taking into account cultural precedent texts.

According to the first approach, precedent texts in scientific discourse are classic works, titles of articles and monographs, textbook illustrations and examples [4]. This approach is in line with the traditional concept of Yu.N. Karaulov, according to which the precedent texts belong to the general culture, in this case, professional culture.

Precedent texts in the form of intertextual inclusions (precedent phenomena in a broader sense) can perform various functions in scientific discourse: password, informative, evaluative, predictive, aesthetic functions. The password function means that references to certain texts can only be understood by members of a given professional group, forming a sense of group identity. The informative function allows the reader to inform the reader about other works in the relevant field of knowledge. The evaluative function of the precedent text gives the author the opportunity to express their attitude towards the described object or phenomenon. The predictive function is realized when the precedent text is in the title or epigraph of a scientific work, preparing the reader to perceive the text. The aesthetic function of precedent texts in scientific discourse is not the leading one and is rather rare. Here we can say that the author turns to a precedent text solely to emphasize the beauty of the wording and the originality of the image [5].

If we extend the functions of precedent texts to the scientific and engineering discourse, we can assume that the main purpose of intertextual inclusions in this case is to inform the reader about other scientific works on this topic, i.e. informative function. In addition, the precedent texts also perform a password function, allowing authors to identify and present themselves as members of the professional community.

A different approach to the types and functions of precedence in a scientific text is proposed by E.A. Bazhenova. A precedent text in scientific discourse is a "quantum of old knowledge" in the form of a term-type concept comprising an author's name and a noun with a general scientific meaning (for example, the Pythagorean theorem, Fourier's law, Planck's constant, etc.) The use of precedent texts in this case is purely pragmatic, therefore, the cultural precedent texts of this type are almost non-existent. The precedent character of these texts is proved by the fact that they are not marked as quotations, are not marked with footnotes and are not referenced. Precedent texts of this type perform epistemological (a tool for systematizing, storing and transferring scientific knowledge), pragmatic (saving text space), social (indicator of belonging to the scientific community) functions [6].

The third approach to the professional aspect of precedent texts is enabled by the research by N.V. Petrova and S.V. Myskin. Based on the levels of precedence identified by Yu.E. Prokhorov, N.V. Petrova came to the conclusion that there are sufficient grounds to classify the texts that are significant for a scientist or a professional and are reflected in their own works as precedent texts and was the first to suggest using the definition of "precedent" for any text which is referenced in the analyzed text, and consequently for any research paper [7]. S.V. Myskin made a thorough analysis of the significance of PPT for a linguistic professional persona. Precedent texts play a key role in shaping the idea of professional activity, since they are the basis of specialized messages that condition and actualize such activity. A key feature of professional texts is that they are intended not only for reading, but also for the subsequent transfer of their content in the language and implementation in objective activity [8].

In his seminal 1965 work, Networks of Scientific Papers, Derek Price shows that a huge body of contemporary scientific literature is "connected" with a small number of earlier works. The author calls such selected texts "active research front", which in some way correlates with the notion of precedence. In addition, Price explains the withdrawal of scientific works from active citation by two opposite reasons: firstly, the decrease in the significance of scientific research, and secondly, such complete inclusion in the scientific context that the need for citation disappears and "obliteration by incorporation" occurs [9].

Thus, we can distinguish the following types of precedent texts in scientific and engineering discourse: classic scientific precedent texts, term-type precedent texts, professional precedent texts. In the English-language scientific discourse, we also note the category of precedence in the form of classic precedent texts and professional precedent texts. From our point of view, professional precedent texts have the greatest potential in English as a second language education, since their role in the actualization of professional consciousness makes them objectively significant for specialists, which undoubtedly increases the motivation for mastering a foreign language [10]. In addition, the absence of strict requirements for textbook and their fundamental nature (in contrast to the texts of the first type) makes it possible to select a larger repertoire of texts for teaching reading. The difficulty with texts of the first type (classic PTs) is that in order to select them, an English teacher requires a level of highly specialized knowledge comparable to the level of a scientist in a given subject area, or requires a high level of foreign language competence from specialized specialists who are potentially able to take on the role of a consultant. We do not consider precedent texts of the second type as the content of teaching English as a second language to future aerospace engineers, due to the low teaching potential of these texts: in most cases they are international and can be classified as eponyms—terms based on proper names.

3 Specific features of professional precedent texts in the aerospace industry

Let us consider the specifics of precedent in scientific and engineering discourse in the aerospace industry. Based on the fact that precedent texts in scientific and engineering discourse are manifested in the form of citations in new texts, we can assume the relationship between the presence of source texts in the space of scientific and technical discourse and their precedent significance. One of the famous authors in the field of scientometrics V.S. Lazarev notes that the value of scientific work can be assessed only based on its use, and the only tool by which this parameter is measured is citation [11].

After studying the citation patterns of a huge corpus of scientific articles, some authors suggest the citation half-life as the determining criterion of the popularity of an article. An article's citation half-life is the period of time during which a publication receives half of its citations. This construct is used by the publishers of scientific journals in order to estimate the length of the embargo, that is, the prohibition of the author to publish the scientific work in the open access. The open publishing embargo allows publishers to maintain a monopoly on the scientific publishing market without losing their profits. However, studies of citation half-lives can help in understanding how scientific texts function and interact.

Davis and Cochran analyzed about 13.5 thousand scientific journals published by Thomson Reuters for the period from 1997 to 2013. The results of the analysis show that the average half-life of a journal's citation is 6.5 years and annually increases by 0.13 years. As a noteworthy trend, it was noted that, despite the increase in the cited half-life in most cases (the authors explain this by the increasingly fundamental character of applied research), scientific and engineering journals have a shorter cited half-life, which, moreover, tends to further decrease. However, the data on journals of some scientific and engineering topics (table 1), show that the cited half-period for scientific articles on aerospace topics is more than 8 years and has a slight tendency to increase [12].

Table 1. Cited half-life for engineering articles (from [Davis, 2015])

Category	Number of journals analyzed	Mean cited half-life	Growth per annum
Engineering, aerospace	315	8.27	0.09
Engineering, industrial	366	7.81	0.13
Engineering, manufacturing	288	6.60	0.13
Engineering, mechanical	1155	7.36	0.05
Engineering, multidisciplinary	608	7.71	−0.05

The study of the "age" of the cited articles in the field of engineering and technology showed that the spread of Web-based technologies had led to greater accessibility for the researcher not only of the latest, but also of older scientific publications. This trend applies mainly to computer science and economics. In scientific and technical publications in 2013, the number of cited scientific articles over 10 years in comparison with 1990 increased by only 3% and amounted to 34% [13]. Thus, we can conclude that the "usefulness period" of scientific and engineering articles is becoming longer due to modern information technologies, and the period of 5 years, established by some publications as a criterion for the relevance of a scientific source, is unreasonably short even for rapidly developing areas of technology.

A number of English-language studies devoted to the sources used in Master's and PhD dissertations on engineering topics can give an idea of which texts are precedent for postgraduate students of foreign universities. The research carried out by the Library of New Mexico State University focused on the research papers underlying the PhD theses of graduate students in engineering. The researchers examined the structure of bibliography for three time periods; the first segment belonged to the pre-Internet era (1989–1991), the second to the transition period (1999–2001), and the third to the era of the accessible Internet (2009–2011). An analysis of the sources showed that free access to the Internet led to a twofold (compared to the pre-Internet era) increase in the number of links to scientific journal articles, the number of links to network resources also increased, as expected, the number of links to books decreased by almost 3 times. As a result, the structure of the list of references in technical theses is as follows: scientific articles—70%, books—10%, conference materials—9%, websites—3%, official documents and state standards—3%, dissertations—1%, other—2% [14].

4 Criteria for selecting professional precedent texts in Master for Engineering programs

Based on the argumentation of S.V. Myskin concerning the mutual transitions between the actual professional text and the original text (i.e. precedent text), namely, that the precedent text and the new text created on its basis should coincide not only lexically, but also structurally [8], we come to the conclusion that the PPT, which are presented to future specialists, should belong to the same types as the texts that they themselves have to create in the course of their educational or professional activities. We distinguish the following types of English-language professional case texts required by a graduate student of a technical university:

- review articles of a general nature in the specialty;
- review articles on the research topic;
- original research articles on the topic of research;
- patents;
- instructions and user manual for machinery and equipment.

Taking into account the individual and social levels of precedence of a professional linguistic persona, we propose the following criteria for the selection of professional precedent texts for reading in English:

- relevance of the specialty and the topic of scientific research (name, keywords, abstract);
- citation index (for scientific articles, monographs, books);
- scientific authority and importance of authors (H-index, number of publications, affiliation);
- the authority and quality of a scientific publication (impact factor, quartile);
- year of publication (on average, not older than 10 years, an exception can be made for fundamental works in a certain area of knowledge).

The analysis of scientific research on the problem of precedent texts made it possible to identify the texts with highest potential from the point of view of profession oriented teaching of a foreign language. These are scientific and engineering texts that are significant for a particular researcher and contribute to the actualization of professional consciousness and a professional linguistic persona. The main genre of professional precedent texts for Master's degree students in engineering disciplines in general and aerospace in particular are original research articles in peer-reviewed journals, which they use when preparing their own publications and Master's degree thesis. The selection criteria for English-language PPT make it possible to individualize the process of teaching undergraduates in profession oriented reading in English. The lack of research on such a significant topic demonstrates the urgent need to develop a methodology for teaching PPT reading in English.

In conclusion, it should be noted that the international nature of modern science and technology requires a researcher and engineer to be actively involved in scientific and professional communication in a global context. In order to speak the same language with the world community, it is necessary to rely on a common body of knowledge, which is contained in professional precedent texts.

References

[1] Yu.N. Karaulov, *Russian language and language personality* (URSS, Moscow, 2010)
[2] G.G. Slyshkin, *From text to symbol: linguistic and cultural concepts of precedent texts in consciousness and discourse* (Academia, Moscow, 2000)
[3] Yu.E. Prokhorov, *Reality. Text. Discourse* (Flinta, Nauka, Moscow, 2004)
[4] V.I. Karasik, *Language circle: personality, concepts, discourse* (Peremena, Volgograd, 2002)
[5] O.B. Adaeva, Philology. Theory & Practice, **7(18):I**, 22 (2012)
[6] Ye.A. Bazhenova, Perm University Herald. Russian and Foreign Philology, **3**, 32 (2010)
[7] N.V. Petrova, Vestnik IGLU, **2(10)**, 176 (2010)

[8] S.V. Myskin, Philology. Theory & Practice, **10(40):III**, 134 (2014)
[9] D.J. Price, Science, **149(3683)**, 510 (1965)
[10] G.V. Sorokovykh, *Increasing motivation to learn a foreign language*, Proceedings of International scientific and practical conference of teachers, postgraduates, students and practitioners "Actual problems of social stratification and transformation in modern conditions", 16 October 2020, Penza, Russia, 452 (2020)
[11] V.S. Lazarev, International Information Forum, **42(1)**, 3 (2017)
[12] P.M. Davis, A. Cochran, arXiv:1504.07479 (2015)
[13] A. Verstak, A. Acharya, H. Suzuki, S. Henderson, M. Iakhiaev, C.C. Yu Lin, N. Shetty, arXiv:1411.0275 (2014)
[14] P.C. Johnson, Collection Building, **33(1)**, 25 (2014)

Methodology for the formation of an individual approach in the training of targeted highly qualified specialists at the request of the aviation industry

Denis Shavelkin[1] and *Vladimir* Tereshonkov[1,*]

[1]Department 101 "Design and Certification Aviation Technology", Moscow Aviation Institute (National Research University), 4, Volokolamskoe shosse, Moscow, 125993, Russia

Abstract. Modern higher education is a rather complex process that requires a special methodology for building the process of obtaining education by future specialists. In the aerospace industry, to simplify the process of searching for future employees of enterprises among applicants/students, a tool such as targeted training is used. At the same time, it does not guarantee the effectiveness of training a student, due to the linearity of the process of targeted learning. In addition, inertia is another important problem in targeted training. This paper presents a methodology for implementing individualization of the training of target students, which will solve the problems described above. Enterprises classify targeted queries into three vectors. The first vector covers further vocational education (FVE). It is associated with specific qualifications required by the employer and industry. The second vector is the classical scheme for preparing students. According to this scheme, training takes place according to the list of educational programs implemented by the university. But at the same time, this vector has the largest number of problems. The third vector is full-cycle training. The main aspect is the application of individualization of the learning process and individual tutoring.

1 Introduction

The modern aircraft and space industry needs a constant supply of new personnel. As part of the growing competition in the market of military-industrial products, enterprises are forced to equip their design and production units with personnel capable of modernizing existing and creating new models of military equipment.

Due to the increase in the working capacity of the industry and the long-term stagnation of personnel, a significant issue arises related to the shortage of qualified personnel. Selection and training of future employees is a reasonable measure of the employer to ensure its production. One way or another, the functioning of authorities and organizations always depends on people: even automated production requires human management and coordination. In this regard, the urgent need of employers for personnel entails an interest in their independent training and "booking" of highly qualified specialists at the formative stage.

*e-mail: tereshonkovva@mai.ru

2 Solving personnel problems

Most enterprises solve the issue of new employees with the help of targeted programs. The classical target program has the following model: an applicant enters into an agreement with an enterprise before entering the institute, enters the institute and after training goes to work for this enterprise. This model is not always beneficial for both the enterprise and the student. Let's consider the problems that may be associated with such a model of targeted training of personnel for industry.

The main problem is the linearity of the process of working with a young frame. The industry pays for an applicant, and receives an employee in 4–5.5 years. In this linear process, there are no stages of a rapid learning process and rapid introduction of new personnel in the industry [1]. In this case, quite a large part of the time is spent on mastering the missing professional competencies, general cultural and universal competencies by the listener. As a result, the industry has to wait until the student fully masters professional competencies, graduates from the institute and comes to work.

Another important problem is a certain conservatism inherent in the classical model of higher education. Conservatism, of course, allows us to preserve the traditions of scientific and educational schools, contributing to the preservation of the best practices of fundamental training of students. This by no means indicates stagnation in higher education: the world's leading and Russian universities are constantly testing new methods and educational technologies, introducing only time-tested and experienced innovations into the main educational programs. Such an approach slows down the process of preparing the target student.

Let's assume a request comes to an educational organization from 5 different enterprises, which on average need no more than 8–10 people per year, who need special knowledge of software. However, the enterprise is not ready to additionally finance and unify its requirements. There is a certain problem in the requirements from enterprises due to the fact that different enterprises need different software products. It is not profitable for the Institute to open a new program/discipline for such a small number of students as part of the usual training of students, so often such industry requests remain fully unsatisfied.

For coping with some of the main problems of the industry in training targeted personnel there are certain solutions that can be used as part of the educational process. These solutions are used in the Moscow Aviation Institute (National Research University) (MAI), in particular, at the Institute No. 1 "Aviation Technology". In this article, we will consider the methodology for implementing individualization to target students' training.

3 Implementation of the target contract

The targeted training of students is significant for all three participants in the process: industry, institute and student. All three participants should be interested in this process. If one of the participants has no interest, then this linked chain is lost, and we return to the problems described above.

The requirement for a specialist, first of all, should come from an enterprise. This requirement for targeted training should take into account the strategic development of the enterprise and the industry as a whole. Based on its strategic path, the organization forms a list of competencies that they want to see in their potential employees. It is noteworthy that the list may not include oly professional competencies but also requirements for Soft Skill, IT, etc. competencies. But the company selects its requirements for competencies for jobs where graduates of targeted programs will be arranged in the future. It is important to understand the following: the ratio of competencies of the target student should be correlated with their salary. Also, the company must declare the required number of applicants for targeted training.

After the formulated requirements from the enterprise have been received by the institute, the educational organization conducts analytical work on the declared competencies. Generally such target requests of an enterprise are classified according to three vectors.

The first vector is the training of students/employees within the framework of program of additional professional education (PAPE). Training under the programs of additional professional education is to a greater extent related to specific qualification requirements required by the employer and industry. This vector can be considered the fastest in the stage of employee training. Professional retraining can take from several months to several years, depending on the complexity of the program. Another advantage of PAPE is that it can be carried out according to any competencies necessary for the future job [2]. But there are also significant drawbacks, for example, this education is ordered from the institute for separate financing from the enterprise.

The systematic approach to PAPE requires ensuring quality guarantees of additional professional education, the formation of a regulatory framework of methodological materials, a certification system and qualifications based on qualification frameworks, professional standards and other qualification requirements. At the same time, it is desirable to organize a system of advanced training and retraining of engineering and technical workers without interrupting them from production processes. This problem can be solved by using distance learning technologies as part of the educational process [3, 4]. But it is notable that the elements of the teacher's contact work with the listener are an obligatory component of the learning process.

There are a number of necessary steps that need to be taken to implement this training of targeted industry personnel. After receiving the list of required competencies, it is necessary to agree on the content of the program, approve the educational plan and enroll students in the course. After the course is listened to, it is necessary to prove the competencies mastered in the course, usually it is done through an exam, an interview, or a practical test.

The second vector is the classical scheme of student training. The university has a list of implemented educational programs. As described above, the program has the main disadvantages: the time of preparation of students and the lack of variability in preparation. The applicant enters into a contract with the target company for a specific specialty and chooses the institute in which such a specialty is implemented. After entering the institute, the target student learns competencies on an equal basis with all students. The only difference between the target students and ordinary students is that target students from the first to the last year undergo practical training at the enterprise and write diplomas on topics that will be of interest to the enterprise. Sometimes, after the 3rd year, an enterprise hires their students, but this method does not guarantee the most important thing: the provision of the employer with the specialist he needs. A specialist noticed by the employer, even with whom a contract has been concluded, may be lured away by competitors, or may not fully master professional competencies and be expelled.

The above problems can be solved with the help of the third vector—full-cycle training, in which the future specialist is actually assigned to the employer. MAI has been implementing one of the largest targeted training projects for a state corporation since 2020. The main idea of full-cycle training is as follows: the immersion of students into the working atmosphere of industry from the 1st year, the use of individualization in the learning process, the presence of individual tutoring and the training of highly qualified personnel. Let's look at these possibilities of preparation and ways to formalize such targeted contracts in more detail.

4 Essence of the target contract

By virtue of a target contract, a state or municipal body, organization, enterprise and educational institution are obliged to provide social support measures to a citizen, organize educational, industrial and pre-graduate practice, as well as provide employment for a citizen, and a citizen, in turn, is obliged to master the curriculum, undergo educational, industrial and pre-graduate practice, and after graduation work in the organization for a certain period of time. The responsibility of the parties for the violation of the contract is manifested in the fact that the citizen reimburses the body or organization for the costs of providing social support measures, as well as a two-fold fine in relation to such expenses. In case of violation of the contract by the body or organization, a citizen is paid a double fine in relation to the social support measures provided.

The link between the student contract and the contract on target training is that the contract on target training can be regarded as a type of student contract due to their similarity. Such systematization is necessary to introduce certainty on some issues. A distinctive feature of the contract on targeted training from the traditional student contract is that the employer, in addition to the obligation to employ, also has the obligation to provide social support to the student (in the form of a "scholarship", the nature of which differs from the state scholarship, or other social support measures). The differences can be traced in two aspects: in the parties to the contract and liability. On the part of the employer, any organization can be a party to the student contract, except for state organizations that can act as a party to the contract on targeted training (since there is a special regulatory regime for them).

Responsibility under the contract on targeted training has been increased: a citizen who has not fulfilled his/her obligations actually reimburses three times the amount of social support measures provided to him, and under the student contract—in a single amount. At the same time, if the employer—party to the contract on targeted training fails to fulfill his/her duties, he is obliged to pay a fine to the citizen, twice in relation to the social support measures paid [5]. In case of violation of the contract, there is equal mutual responsibility—eventually one of the parties will receive three-fold measures of social support. Responsibility under the apprenticeship contract, as previously written, performs a compensatory function and is associated only with reimbursing the employer for his/her expenses.

As soon as the applicant enters the institute, a lot of work begins with him. First of all, each student has his/her own supervisor attached to him, who explains the whole system in detail. Right from the first semester, these students' education is built up several levels higher [6]. In the educational process, they are taught English at a high level with the number of hours 2 times more than ordinary students. One part is implemented as part of the curriculum, the second as part of additional classes, for which the customer pays. In addition, students participate in international conferences in English, first as listeners, and then as participants.

Also, from the first semester, students have a whole day in the middle of the week, during which they get acquainted with the company, and after some time they are officially employed and start working there [7]. In the first semester, students begin to study work with specialized IT products, as well as in-depth acquaintance with modern technologies in the field of robotics, composite materials, 3D printing, etc. Students undergo in-depth training that stretches over several semesters in programming in C++, Java or Python. In addition, they master computer-aided design systems (CAD), engineering analysis (CAE), technological preparation of production (CAPP), life cycle management systems (PLM) at a professional level. This training is also well suited for distance learning. The process of distance learning has already been sufficiently developed and there are many software products used in conducting such classes.

After the first year of study, the student not only passes an internship, but is also enrolled in a working profession, such as an aircraft mechanic, a CNC machine operator, etc. These additional skills allow the target student to participate in the work process in the specialty field. At the same time, the tutor suggests where you can apply your knowledge and additionally learn something new. Such additional courses, the development of individual training for students and tutoring allows to build a trajectory of a highly qualified specialist for industry, who from the first months of study will immerse himself in the working structure and improve his/her knowledge under the employer's order.

In subsequent courses, the student also builds an individual structure. Depending on the specific requirements of the customer, academic mobility from 6 to 12 months to universities in other countries is available for students. For academic mobility, the company can send students together with the institute to universities such as: Mendoza University, Buenos Aires Civil Institute of Space Technology, Bandung Institute of Technology (ITB), Shanghai Transport University, Beihan University, Beijing Polytechnic Institute, Technical University of Vienna, Technical University of Munich, University of Applied Sciences Ingolstadt, University of Applied Sciences Heilbronn, University of Stuttgart, Polytechnic University of Madrid. Graduate School of Aeronautics and Space Research, Polytechnic University of Turin, ECAM LYON, ISAE–SUPAERO, IPSA, ENAC.

5 Advanced MAI programs

It is important for the company to prepare not only a good specialist with a high knowledge of professional competencies, but also a potential manager. Therefore, the Institute offers these students to undergo professional retraining within the framework of the "School of Service" and "School of Management".

The School of Service is the first aviation service program in Russia. After-sales service of aviation equipment is one of the promising areas in the industry, located at the intersection of aircraft construction, economics and management. The Service School is attended by organizers from Moscow, experts from Italy, France, Japan and participants from all over Russia. There are 5 thematic blocks in this school: the current state of MRO in Russian aviation, the business model of the service and the contracting system, the design of the aircraft in the logic of the cost of ownership, the operational infrastructure: MTO and MRO operational infrastructure.

The MAI School of Management creates an environment for the development of talents and provides successful engineers with managerial competencies as well. The School of Management was created for the personnel reserve of high-tech corporations, talented students and employees of MAI, who are ready to develop their native institute and implement promising projects. The project was launched jointly with the largest corporations—the United Aircraft Corporation (UAC), the United Engine Corporation (UEC) and the United Shipbuilding Corporation (USC).

At the Moscow Aviation Institute, there is another stage of training highly qualified specialists for the requirements of industry—a targeted master's degree. The Master's degree program has a certain flexibility in implementation and, as a result, is in demand in a competitive market. Enterprises can send students to study under existing programs or sign a contract for the development of a completely new program. As a rule, to such programs companies send employees who already have experience working at this enterprise. Also, under the requirements of the customer, it is possible to obtain a specialist with a double diploma within the framework of the master's degree, while he will receive one diploma at MAI, and another at Shanghai Transport University (Shanghai Jiao Tong University). The MAI—SJTU Joint Institute is a unique educational and scientific environment in which students of the two countries study in joint groups throughout the entire period of study, which allows them to achieve high quality results in education and exchange their experience [8]. These programs are a direct example of the theory of global education and have no analogues in Russia and abroad.

6 Conclusion

The purpose of full-cycle training is to educate a new generation of engineers with a set of "digital" competencies, fully ready to work in modern aircraft construction. The training program is designed taking into account the real demands of the industry and helps to build a professional trajectory of future specialists from the first year. Students will be maximally involved in the current production processes of aviation cluster companies. The development of such programs is a condition for the competitiveness of Russian industry in the future. To continue studying under the program, students must annually confirm the achievement of targets. Upon graduation, enterprises guarantee employment of graduates who have achieved the goals set by the employer.

The ways of introducing individual training of highly qualified specialists for the needs of industry were proposed above. The requirements for a student and a future employee of the industry are put forward by the industry itself together with the institute, thereby forming new highly qualified personnel.

References

[1] E.G. Elina, V.M. Anikin, *Purpose Education: Social Risks and their Overcoming*, Izv. Sarat. Univ. Sociology. Politology, **18**, 373 (2018)

[2] R.R. Anamova, L.V. Bykov, D.A. Kozorez, *Aircraft Industry Staff Retraining as a Part of Vocational Education in the Russian Federation*, TEM J., **8(3)**, 978 (2019)

[3] E.A. Volzhenina, V.A. Kaigorodova, D.S. Shavelkin, *Methodology for Conducting Classes in a Remote Learning Environment*, Modern Pedagogical Education, **10**, 58 (2021)

[4] V.E. Chernoskutov, *Problems of Purposeful Contract Education of Governance Personnel*, Problems of Governance. Scientific and Practical Journal, **7(10)**, 40 (2014)

[5] G. Gilpin, M. Kofoed, *Employer-Sponsored Education Assistance and Graduate Program Choice, Cost, and Finance*, Res. High. Educ., **61**, 431 (2020)

[6] J. Glaesser, *Competence in educational theory and practice: a critical discussion*, Oxf. Rev. Educ., **45**, 70 (2019)

[7] P.G. Filippov, A.S. Tutushkin, *The formation of the Continuous Education System in the Rostec State Corporation*, Innovations, **12(182)**, 24 (2013)

[8] I.A. Konstantinov, V.A. Goncharova, *Competition Model in the Russian Higher Education System*, Education Management Review, **3**, 111 (2021)

Virtual simulators application in the foreign language studying as the factor of students motivating

Elmira Shefieva[1,*], *Oksana* Simonova[1], and *Olga* Bessarabova[1]

[1]Rostov State Transport University, 2, Rostovskogo Strelkovogo Polka Narodnogo Opolchenia sq., Rostov-on-Don, 344038, Russia

Abstract. Our research paper draws attention to the problem of exploiting of the Virtual Reality simulations in studying foreign languages and we also attempt to authenticate the positive impact of employing this technology in forming an educational motivation of the students. The urgency of the investigations was promoted by the utter spreading of the immersive technologies interwoven into all spheres of our daily life. The aforementioned factors arise the significance of changing the pedagogical paradigms and doctrines to be up-to-date with the things change. Our study detected that being a complex phenomenon motivation plays a significant role in the educational process; the experimental cohort comprised 64 first year students of the of the Humanities Department, of the Hotel Business and Tourism Business specialties of the Rostov State Transport University who were asked a list of questions connected with the motivational factors; we chose a "Field Trip" simulation as a learning tool for our experiment and then statistically validated the fact of educational motivation rise after the incorporating of the aforementioned technique into the procedure of the studying of a foreign language.

1 Introduction

Twenty-first century can be correctly characterized as the digital age because of the deep immersion of the people throughout the world into the smart devices exploiting.

Figure 1. Adults' Emotional Attachments to the Gadgets Chart [1]

*e-mail: sherifovna@mail.ru

Within the past, fewer years having smartphone in your pocket has become a commonplace helping ordinary citizens carry out mundane things. Modern technologies created a digital revolution offering us numerous benefits and changing our life paradigm simultaneously. It appears to be that we have a voracious appetite for information in our fingers. The Pantas and Ting Sutardja Center for Entrepreneurship & Technology [2] highlighted the results of Dr. Veronika Konok and her team those cite that even responsible adults are attached so much to their smart gadgets (figure 1). The same institution outlines that the consumption of the digital devices has become disproportionate withing the last decade. We can observe the growth comprising 300% in the period between 2011 and 2016 and 23% growth by 2020. The statistics numbers are widespread all over the world e.g.: Felix Richeter, the data journalist from Statistica.com interprets the results of the Internet utilizing by the American teenagers in the form of the diagram that indicates that almost three thirds of the youth spent almost all the day being online withing the frame of that or this social net (figure 2).

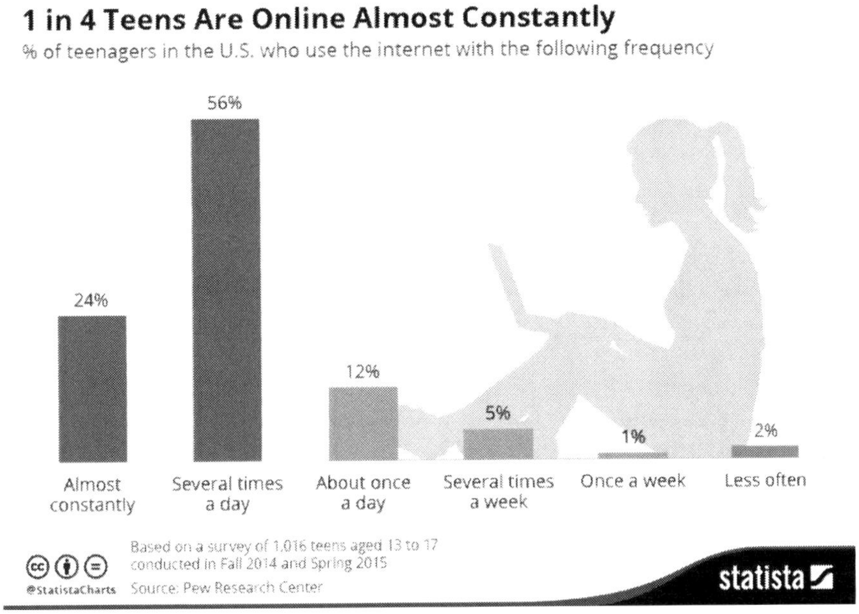

Figure 2. American Teens Internet Frequency Use Chart [3]

Bearing in mind the aforementioned statistic data we should recognize the urgency of the modifying of the classical teaching methods and adopt them to be relevant to the time challenges and eventually to help the students grasp the material better in connection with the foreign languages study.

This section presents a recent focus on the aforementioned topic by means of the brief analysis of the literature about the utilizing of the innovative digital technologies employed by the educators into the studying process considering the motivation aspects. We must mention that motivation plays a vital role in studying in general and in studying foreign languages in particular. The academic community has extensively explored the topic of using the innovative methods employed by the educators with regard to studying foreign languages and we can posit that much is known and at the same time there are certain gaps and much space for investigations in this area of study, highlighting the correlation between the Virtual Reality (hereby simply referred to as VR) incorporating into the teaching process and students' motivating for foreign languages mastering.

F.G.E. Fandino et al. [4] highlights that motivation turns out to be one of the most significant factors in studying a foreign language; he supports the idea of Woon et al. [5] that motivation is a mixed process including many spheres of the human behavior such as: biological aspects, cognitive aspects, aspects connecting with the action taking; the findings of the team of the scientists discovers that motivation in influenced by many external elements. Jafar Pourfeiz [6] from the Hacettepe University, Ankara explores the correlation between the academic motivation and the foreign languages studying interpreting the results of the experiment as the reciprocal interdependence of the abovementioned constituents and highlighting a special role of the estimating components of the inclination to foreign language study. Ediz Tuncel et al. [7] findings on this topic demonstrates that there is a certain contradiction in positive performance and negative motivation in studying a foreign language in the experimental target group of students. W.L. Quint Oga-Baldwin and Luke K. Frye [8] detected interesting facts in the framework of their study, they discovered that gender influences the motivation to study foreign languages positing the idea that female teens are more motivated to learn foreign languages comparing with male ones. W.L. Quint Oga-Baldwin also states the difference of Western and Asian type of motivation in learning a foreign language in favor of Asian one. Mohialdeen Alotumi [9] supports the same ideas of gender influence on the motivation of study a foreign language, his findings also prove the fact that girls in colleges manifest more motivation than boys being the participants of the experiment in target groups on the study.

Meeting today's requirements educators all over the planet find the idea of transforming the teachers' strategy as topic significant and to support these trends many scientists explore the innovative methods of teaching including VR which in other words is called simulation and presents the digital technology facilitating us with the immersion into the imaginary world. Merriam Webster's dictionary gives the following definition of this phenomenon as "an artificial environment which is experienced through sensory stimuli (such as sights and sounds) provided by a computer and in which one's actions partially determine what happens in the environment" [10].

This method is relatively easy to build in to the education process due to the wide spread of the digital technologies however if we evaluate the width of spreading of the VR technologies in a foreign language study (hereby simply referred to as FLS) we can recognize a small portion of scientific studies connecting this very subject. The major part of all investigations on the aforementioned theme is dedicated to the study of simulations in medicine study management and law. Thus, Ram Nataraja [11] examines the simulations in the sphere of laparoscopic skills education process; Madeline C. Mier et al. [12] explores the simulations in the field of emergent pediatrics; Michaela Jánošiková [13] consider simulation in crisis management education in the group of Visegrad countries. At the same time a few of the educators still set up a question of VR techniques in SFL, e.g., Tynkaliuk Oksana V. and Semeriak Iryna Z. [14] survey of the PC technologies and Internet facilities in connection with SFL. This scientific tandem offers such techniques as online chat training and game playing simulation methods for SFL. We also want to mention the academic paper of T. Isaeva et al. [15] which points out the pedagogical aspects of simulation technology exploiting in connection with the SFL.

Bearing in mind all the aforementioned points our team proposes that the utilizing of the VR applications within the frames of SFL can contribute to the formation of the students motivating; the objectives of our study paper is to identify the positive effect of the incorporating of the VR technologies into the education process for forming students' motivation for SFL.

The studies have shown the lack of the methodological and pedagogical materials dedicated to the using of the VR applications in studying foreign languages in terms of motivation forming.

2 Materials and methods

Describing the prior research, we can deduce that all the VR applications shown in table 1 have their strong and weak points; we summed up the advantages and disadvantages of the described ones below.

Table 1. VR Applications

Application	Description	Advantages	Disadvantages
360° videos: Expeditions	Case online & downloaded videos	Possibility to visit different places, developing speaking competencies	Needs the Internet, Internet Limitations
Google Arts & Culture	Online Art content from more than 2000 museums and archives	Possibility to widen the culture study knowledge, developing speaking skills, creating friendly atmosphere	Needs the Internet, emotional lack
Google Cardboard Box	Immersive technology with VR viewers	Affords to widen the speaking and writing skills	Cannot be online, unsuitable for group work
National Gallery	Online Gallery tours	Possibility to watch word masterpieces speaking skills improving	Needs the Internet, emotional lack
"Fields Trip"	A role play modified with VR technologies	A big variety of cases developing speaking skills	Needs the Internet

Many practicing educators such as Paul Driver mention that we have no need to invent the wheel again in the sphere of teaching methods in connection with the changing the education environment evoked by the incorporating the digital technologies into the studying process and we still can apply the traditional teaching approach such as role play called "Field trips" upgraded by means of digital technologies [16]. The role play game implies many modifications such as: at the hotel (for the hospitality industry), at the HR department (for the future HR specialists), at the advertising agency (for the PR managers) etc. "Field trips" can help in studying for the future specialists of the tourism and hotel business industries by letting the students travel virtually to different countries and to communicate in different workplace situations, to employ the "Field trips" technique educators apply the "Google expedition" application. Raquel Ribeiro [17] introduces the "Google Cardboard Box", "National Gallery", "Google Arts & Culture", "360 videos" and "Expeditions" applications for using within the frames of FLS, she mentions that the first one (Google Cardboard Box), cannot be used nowadays because of the Covid-19 spread hence she deduces that the other aforementioned four applications can suit best for the purpose of utilization during the remote English classes. All the applications, of course, have their own peculiarities, thus the educator asked children to make their own videos commenting on the seen piece of arts in Instagram; the advantage of this kind of activity was the possibility to narrate stories anonymously which contributes to the creating of the friendly atmosphere during the classes which in its turn rise the motivation for studying. Raquel also gives instructions on using the application called "360° videos", they can be easily found in YouTube with the indication in the left upper angle of the screen with the selected video. One of the interesting variants of incorporating of the "360° videos", is the "Expeditions" containing vast variety of the narrated tours besides this app can be modified by means of not only VR but also AR (augmented reality).

Analyzing the above-mentioned table, we can observe that the all the online applications need to be provided with the high-speed Internet technologies though there are places where such technologies are still not available; educator should be creative and the students should sometimes have sound knowledge of the language to be involved into the VR provided activities, methods and strategies chosen by the teacher should be pedagogically relevant and justified.

The changing educational environment calls for changes in the teaching paradigm to face the time challenges so fast that we should adopt the new pedagogical methods and strategies faster and apply them more frequent in our daily educators' routine.

To describe the analytical scope of the paper we have chosen the "Field Trips" digital role play for our experiment.

The initial sample consisted of the two groups of students comprising 64 participants of the Hotel business and Tourism Business specialties, the Department of Humanities of the Rostov State Transport University.

The average age of the participating students was 18. Females comprised 76% ($n = 48$) and males comprised 35% ($n = 18$) of the contingent, correspondingly. Regarding the year of study all the participants were freshmen. The students could be characterized as having relatively the equal level of knowledge of the Foreign Language (English one).

The experiment was conducted within the frames of one academic year beginning on the 2nd of September 2019 and finishing on the 31st of May, 2020.

We employ the VR role play that is "Field Trips" in a synchronous environment because of the epidemic lockdown and one more fact contributing to our choice is that this VR digital role play best suits for the students to develop speaking skills in the professional sphere for the mentioned specialties of Hotel and Tourism Businesses.

Our digital role play presents a downloaded application for smartphones called "Google Expeditions", as the targeted groups were the future employees of the recreational activity sphere our research team created a game "A tour Guide". The participants were offered to find information according to the chosen tours "Moscow" and "Travelling across Saint Petersburg" and "Lake Baikal" to perform as a tour guide. The "Moscow" virtual 360° video tour consisted of incorporated five videos of Vodovzvodnaya Tower, Savior Tower, Bolshoi Theatre, Statue of Peter the Great and the panoramic video of the Kremlin from Manege Square. The tour "Lake Baikal" comprised eight videos those were: Lake Baikal with Trans-Siberian Train, traffic signs on the frozen lake, Russian Banya on the Ice, The interior of a Small Russian Orthodox Church, Lake Baikal Ice, Lake Baikal Omul Fish Sold in Listvyanka Market, Interior of the Ivolginsky Datsan Buddhist Monastery, Ice Fishing Hole. All the videos contained podcasts; also, the participants were offered to create their own tour about the city of Saint Petersburg called "Travelling across the Saint Petersburg".

To estimate the motivation constituents, we asked the participants to answer the question "How often do want to use VR during the classes?" before and after the experiment and the range of variations were: 1) almost always; 2) frequently; 3) sometimes; 4) rarely; 5) almost never.

We also asked students to answer the questions to evaluate their desire to use the VR within the frame of the SFL classes by means of commenting on the following statements: I would rather exploit the VR because: 1) it's enticing; 2) it's useful for my language proficiency; 3) VR stimulates my individual work. To estimate the answers, we used a 5-point Likert type scale containing the variations from: 1 (strongly disagree); 2 (disagree); 3 (neutral); 4 (agree); 5 (strongly agree).

3 Results

Our findings indicates that before the experiment 23% of the students were eager to use the VR technologies during the classes and after the experiment almost 31% of the participants demonstrated the desire to exploit the VR application for SFL.

Below we can see the pie charts (figures 3 and 4) indicating the difference in numbers before and after the experiment.

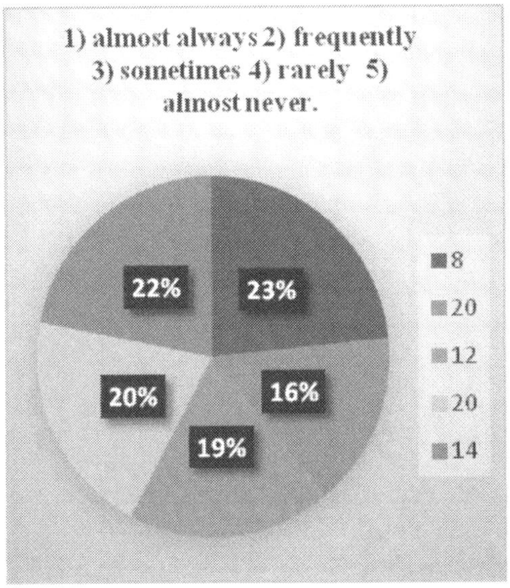

Figure 3. Pie Chart diagram indicating the frequency of VR apps use before the experiment

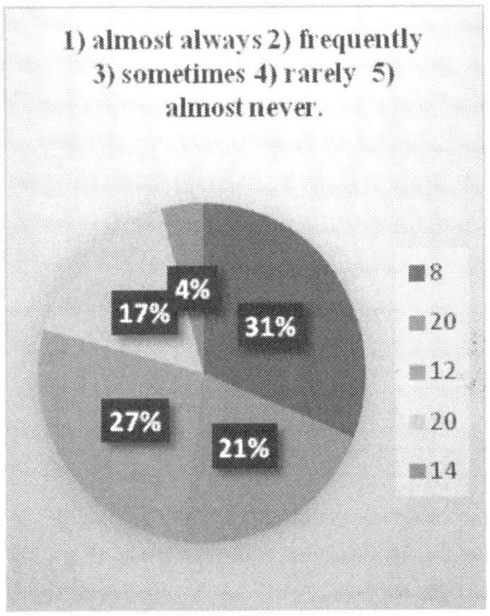

Figure 4. Pie Chart diagram indicating the frequency of VR apps use after the experiment

Analyzing these two pie chart diagrams we can detect the difference in frequency of the use eagerness comprising 8% in favor of exploiting the VR technologies during the study.

As for the questionnaire about the reason of using the VR applications for SFL taken after the experiment, the results were as follows: the entertaining factor was predominant but the proficiency factor also was among the leading factors.

While considering the results of students' motivating for exploiting the VR applications for SFL, we can indicate that entertaining reasoning and convenience are predominating over the other factors influencing students' motivating, at the same time we should highlight the creating of the friendly atmosphere to be among the leading factors that impact on the motivation for study.

We also process the data before and after the experiment on the students' eagerness to use VR applications by means of the statistic method that is *T-students criterion* designed and intended for comparing the average results among small groups within the frames of the experiment and the results detected that the observed differences in the attribute data are statistically sustained and indicate the significant difference because $p < 0.05$.

4 Discussion

The abovementioned results and data support the idea of the prominent American linguist Stephen Krashen [17] who proposed the dependance of the results of studying a foreign language on the mood of the students within the frames of his "Input Hypothesis", indeed the entertaining component of the VR applications technologies contributes much to the motivation, on the other hand the VR utilizing methods follows the modern trends in the society that is the spreading of the omnipresent digital technologies which enter all spheres of our life making the possessing of the smart gadget to be a commonplace and giving birth to insatiable need for information in our minds. At the same time our study paper achieved the to authenticate the firm domination of the motivation factor of the SFL when using VR simulations.

5 Conclusion

This study illustrates the utter urgency of changing the educators' techniques and doctrines under the modern conditions of the fast-changing reality which is deeply involved into the immersive technologies. Our studies promote the wide introducing and incorporating of the VR simulation methods into the educational process and the curriculum on studying foreign languages in particular. Our work has a practical and applicable significance; it's easy to reemploy in any education curriculum. The applying of VR technologies in the educational process though has some gaps and unexplored areas and we understand that it is an open-ended scientific question.

References

[1] Berkely.edu, URL `https://scet.berkeley.edu/wp-content/uploads/Screen-Shot-2018-07-10-at-3.52.31-PM-300x163.png`
[2] A. Indrakanti, M. Chutake, S. Prouty, V. Sundaranatha, V. Koverkathu, *Gadget Addiction*, URL `https://scet.berkeley.edu/reports/gadget-addiction/`
[3] F. Richter, *1 in 4 Teens Are Online Almost Constantly*, URL `https://www.statista.com/chart/3387/internet-use-teenagers/`
[4] F.G.E. Fandiño, L.D. Muñoz, A.J.S. Velandia, Helyon, **5(9)** (2019), DOI: 10.1016/j.heliyon.2019.e02394
[5] W.C. Liu, J.Ch. Keng Wang, R.M. Ryan (eds.), *Building Autonomous Learners. Perspectives from Research and Practice Using Self-Determination Theory* (Springer, 2016)
[6] J. Pourfeiz, Procedia Soc. Behav. Sci., **232**, 668 (2016), DOI: 10.1016/j.sbspro.2016.10.091
[7] E. Tuncel, S. Sadikoglu, K. Memmedova, Procedia Comput. Sci., **102**, 244 (2016), DOI: 10.1016/j.procs.2016.09.397
[8] Q. Olga-Baldwin, L. Fryer, Helyon, **6(5)** (2020), DOI: 10.1016/j.heliyon.2020.e04054
[9] M. Alotumi, Helyon, **7(4)** (2021), DOI: 10.1016/j.heliyon.2021.e06664
[10] Merriam—Webster Dictionary, *Definition of virtual reality*, URL `https://www.merriam-webster.com/dictionary/virtual%20reality`
[11] R. Nataraja, Ultrasound Med. Biol., **45**, 31 (2019), DOI: 10.1016/j.ultrasmedbio.2019.07.510
[12] M.C. Mier, A.A. Nelson, D. Vander Roest, Acad. Pediatr., **20(7)**, 25 (2020), DOI: 10.1016/j.acap.2020.06.074

[13] M. Jánošiková, Transportation Research Procedia, **40**, 1372 (2019), DOI: 10.1016/j.trpro.2019.07.190

[14] O.V. Tynkaliuk, I.Z. Semeriak, Information Technologies and Learning Tools, **67(5)**, 239 (2018), DOI: 10.33407/itlt.v67i5.2272

[15] T. Isaeva, N. Malishevskkaya, E. Goryunova, L. Lazareva, M. Churikov, E3S Web of Conferences, **273(12011)** (2021)

[16] P. Driver, *How can we use Virtual Reality to teach English?*, URL `http://www.cambridge.org/elt/blog/2018/01/25/virtual-reality-teach-english/`

[17] R. Ribeiro, *Virtual reality in remote language teaching*, URL `https://www.cambridge.org/elt/blog/2020/10/27/virtual-reality-in-remote-language-teaching/`

[18] S. Krashen, *Stephen Krashen's Theory of Second Language Acquisition*, URL `https://www.sk.com.br/sk-krash-english.html#:~:text=The%20Input%20hypothesis%20is%20Krashen's,second%20language%20acquisition%20takes%20place.&text=For%20example%2C%20if%20a%20learner,level%20'i%20%2B%201`

The effectiveness of the augmented reality application in foreign language teaching in higher school

Oksana Simonova[1,*] and *Alvina* Kolesnichenko[1]

[1]Rostov State Transport University, 2, Rostovskogo Strelkovogo Polka Narodnogo Opolchenia sq., Rostov-on-Don, 344038, Russia

Abstract. In this study paper we consider the application of the augmented reality for teaching a foreign language and its influence on the education environment throughout the world since educators are trying to find the relevant methods of teaching to better face the time challenges connected with the expansion of the computer technologies. The given paper aims at validating the idea of the effectiveness of the application of the augmented reality techniques in teaching a foreign language at the University. The survey conducted by the authors comprises 64 participants of the Rostov State Transport University of the HR Department; the scrutiny endorses the proposed hypothesis.

1 Introduction

Our society witnesses the expansion of the digital technologies spreading. Due to the nature of this phenomenon and relative ease of use, almost any person can immerse into the world created by the artificial technologies. It is obvious that the digital technologies phenomenon enters all spheres of our life including education, giving rise to changes in methods of teaching, teaching strategies and teaching doctrine in general. We must remind of the fact that digital proficiency nowadays is considered to be the integral part of the prosperous existence of the person in the society. International organization P21 (Partnership for 21st Century skills), which presents the association of the most influential and huge business and science organizations and governmental agencies, includes digital literacy into the list of the skills needed in the 21st century for the individual to be socially adaptive and to function successfully in the environment [1]. New generation colloquially called as zoomers who presents the scope of people born between 1997 and 2010 differs greatly from the previous generations: they are technology-friendly people, deeply involved into the social media life; it means we must bear in mind that the education process must face the challenges of the contemporaneousness to be relevant the general trends in society development. New methods of teaching must support lecturers by means of offering them new variety of training methods and technologies such as augmented reality (hereby simply referred to as AR) that is under the study of our paper.

In this section we will scrutinize the scientists who are involved into the study of the augmented reality. Firstly, we must name the author of the term who is Thomas P. Caudell, a former Boeing engineer who applied the aforementioned scientific know-how to help workers with the cabling in the air plane, since that time this term came into a frequent use for

*e-mail: simnvoks@mail.ru

the describing the technology defined by the D. Roopa et al. [2] as a mixture of the real and computer created universe stuffed with the audio, video, textual and the wide array of other properties for creation a sound educational practice for the user, the authors of the very research paper outline the effectiveness of the abovementioned technology for improving the quality of the educational process. A wide spectrum of scientific papers is dedicated to the study of the augmented reality in medical education.

The following scientists are engaged into the investigations of the medical augmented reality applications: T. Joda et al. [3], Tony Liao et al. [4], Daniel J. Thomas [5] and many others. A special interest for our research paper presents the work of Alexandra Klimova et al. [6] who were examining the existing practices in teaching English using the augmented reality applications. Che Samihah Che Dalim et al. [7] analyses the AR application concerning the teaching English to a non-native children's language learning. Hongyan Bai et al. [8] describes the technology called "English smart classroom" as innovative one in comparison with the traditional class studies. The study paper of Juan Garzón et al. [9] about the interdependence of the pedagogical approach and AR in education also contributes to the methodological base of our research work. Ting-Chia Hsu [10] from National Taiwan Normal University in his research work dedicated to AR asks a question about the importance of the teaching styles in the English practice.

The simplicity of use and relatively inexpensiveness of the AR technologies arise a great interest among the educators and students to such kinds of applications around the world. By the Perkins Coie and the XR Association [11] the percentage of the students asked a question which sphere such kind of technologies as AR is applicable to, gave the answer that was "education"—41% and that was the second place after gaming comprising 61%. Moreover, computer technologies in education create a friendly environment that contributes to the rise of motivation in learning languages [12].

The analysis of the scientific papers on the topic showed that the application of the augmented reality in education and teaching foreign languages in particular is a very urgent problem resulting in changing of the teaching techniques and strategies but still there are some gaps in a foreign language teaching method as a result we posit that the application of the AR tools in teaching a foreign language makes the process of education more effective. In our investigations we apply the following scientific methods: comparative analysis method, descriptive analysis, observation method and mathematic statistic method of scientific research.

2 Methods and materials

We conducted a fortuitous controlled attempt to authenticate the positive impact of the AR application on the teaching foreign languages process. The experiment started on the 2nd of September 2019 and finished in May 2020 comprising one academic year. The measurement and observation took place at the Rostov State Transport University (hereby simply referred to as RSTU), at the Humanities department by the corresponding authors. The selected reference and the control groups are the students of the Human Recourses specialties. The initial sample consisted of 64 participants (that comprised 4 educational groups of students) studying at on department of the RSTU of the Humanities department, females—76% ($n = 48$), males—34% ($n = 21$) of the cohort ($n = 64$), respectively. All the participants were freshmen. All the members of the selected groups could be characterized as having relatively equivalent educational level of English. During the first classes the participants were given explanatory instructions to get familiar with the interface of the two applications those were: Tour Builder &Tour Creator. The experimental and the reference groups received materials containing the indistinguishable subject matter. For analyzing we took the final results on the examining discipline of English Language of the reference and the experimental groups;

for data handling we utilize the mathematical statistic method of data manipulating called U-criterion. The scatterplots illustrate the following correlations between the reference and experimental group; hence, the arguments advance in favor of experimental team indicating $P \leq 0.01$. At considering the results we deduce that the exploiting of the AR applications in teaching a foreign language at the University upgrades the level of the knowledge of the students.

Basing on the results of the statistical analysis we asked students to comment on the list of questions to evaluate the exploiting of the AR technique in the educational process; the participants were offered to esteem the AR classes atmosphere handling the Likert type 5-points scale: 1 point (strongly disagree), 2 points (disagree), 3 points (strongly agree), 5 (not applicable). The learners were given the following statements: 1) I would apply the AR because it's more interesting comparing with the traditional classes; 2) I would apply the AR because it's more convenient; 3) I would apply the AR because the atmosphere during the AR classes is friendlier. Out study highlighted that before the experiment the major part of the students used AR less than 2 hours a week and after the experiment. Judging the diagram (figure 1) we can see that they started to use the AR applications in the educational process more often. The participants outlined that the AR utilizing helped them to better grasp the material and created a friendlier atmosphere during the classes which help to master the language proper and to achieve better results in studying in general.

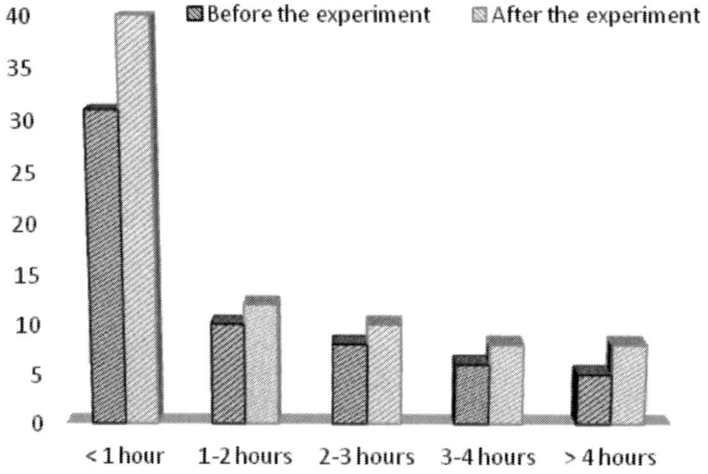

Figure 1. The frequency of the AR application use before and after the experiment

As for the results of the question list the participants in the number of 64 gave 78 answers which indicates different motivation from convenience to a more interesting frame of classes, but the most interesting findings appeared to be the notes about the creating of the safely atmosphere during the classes which contributes to the achieving the better results in studying. The overall goal of this work is to evaluate the validity and the effectiveness of the AR teaching tools in the educational process of teaching a foreign language.

3 Results

Augmented reality is widely used in all branches of science. It finds application in medicine, architecture, aircraft construction as well as train- and car-building, design and many other spheres. Augmented reality (hereby simply referred to as AR) is particularly relevant in the educational process, as it allows attracting students' attention and increasing their motivation in learning a subject. As for teaching a foreign language, this kind of the pedagogical technology helps learners to immerse themselves completely in the language environment. In this case, all the conditions for full immersion including virtual reality are created and students will have an opportunity to talk with a native speaker, listen to the authentic speech or even visit a famous museum in the country of the language being learnt. It became possible due to augmented reality.

There is a great number of different digital technologies being used during practical classes in English at the university that contribute to the development of students' reading, listening, speaking skills. In this article we try to show the advantages of using AR in a foreign language teaching in the higher educational establishment proving the ideas of Vat-U-Lan who considers that those students who use AR in the English course greatly improve their important skills by comparing with others who learn English in a traditional way.

Among numerous educational tools with AR useful in mastering English are those that improve speaking skills. These are social platforms designed to socialize in English with people. They include "Time XR Social AR&VR" where discussions happen in a virtual world and teachers can use it as an alternative to conversation clubs in English lessons. Another way to improve students' conversational speech is "Altspace VR Community" where they will have an opportunity to communicate with real users all over the world in different speech situations. It is a good practice in English especially in free time because its usage helps to develop listening and speaking skills of learners.

Taking into account different platforms aimed at improving English we should mention educational ones. "Class VR" is created especially for the English language teachers but it also includes an independent educational program with prepared lessons. This program enables to discuss real life situations—shopping, study and so on. It can be supplemented with virtual glasses or helmet that allow students watching interactive videos where they act as participants of the process. They can take part in a virtual tour around a British park or celebrate some holiday, for example, Christmas or Easter. "Class VR" offers augmented and mixed reality programs in which you can use textbooks and workbooks. Teachers form a VR room link where a large group can study under the teacher's control during the certain time. Its peculiar feature is that "Class VR" helps students to communicate using interactive means. This course keeps their interest in learning as well as develops communicative competency as one of the most important goals in learning a foreign language.

One of the most exciting educational platforms for learning English is "Engage". It is produced by the company "Immersive VR Education" that makes it possible to use virtual and augmented reality in the educational process giving high schools all tools for creating the content necessary for virtual classes. The free educational platform "Engage" represents virtual space for conducting practical classes in groups or even individually by teachers, for meeting, discussing or giving presentations in English. In addition to this, students can have experiences of virtual reality using such products, as "Apollo 11 HD", "1943: Berlin Blitz" and "Titanic VR" that allow them to land on the Moon, visit Berlin in 1943 or explore the wrecked steamboat, get acquainted with archival audio and video recordings including the BBC archive and feel the atmosphere of real actions. On the one hand, this educational experience will leave an indelible impression on learners, on the other hand, it will promote to language skills development of students.

Another educational course called "Virtual Speech" combines both traditional means of English teaching and modern technologies of virtual reality. This course consists of a great deal of programs including online lessons: reading, translating, work with texts, listening practice and watching films. In addition, students can practice their speech skills in virtual conferences, discussions, performances. One of the drawbacks is that "Virtual Speech" is not free. The main emphasis is made on Business English and this course is useful for the students of economic profile of training as students master business vocabulary, various economic topics and the skills of public speaking so essential for their future professional activity.

Augmented reality in practical classes makes learning English exciting and informative. It allows students to develop their abilities in independent research activities that is essential in a higher educational establishment. One of the ways to improve these skills is to use special platforms for creating presentations with 3D models. "Jig Space" gives an access to different presentations and shows arrangement of complicated mechanisms or objects in all details in the form of three-dimensional instructions, for instance, the layers of the Earth or some furniture, engines, etc. Students can create their own presentations with the help of "Jig Space" as well as "Timeline JS" and make the studied subjects real. The later of the mentioned is intended for visualizing chronological events. This service has wide opportunities because creating their presentations students insert texts from various sources, media-content, video- and audio-materials, web-documents, maps, etc. In this way, learners show their creative potential, abilities to self-study, analysing materials and good knowledge of English.

The services "Tour Builder" and "Tour Creator" have practically the same function. They deal with the creation of own tours across a country, a city or simply virtual excursions in any museum. The work with these services is both spectacular and fabulous because they allow students to find themselves in a virtual city or a famous museum that seem to be realistic. Besides, it reveals a person's creativity and contributes to developing speaking skills. Being a creator and a guide of his own tour, a student can apply all his knowledge of English in the process of communication with a group of visitors.

It is also necessary to pay attention to modern applications that English teachers can effectively use in their practice. Application programs create such an atmosphere that is aimed to attract students with visually bright and 3D-moving models. These programs make them interact with the material and be engaged in learning a foreign language. Among a great variety of applications with the usage of augmented reality in the English language teaching, we should mention the following: "Mondly", "Rumii", "Mind Map AR", "Civilisations AR", "ARki" and "Smart Reality".

The educational application "Mondly" is intended for learning all aspects of the English language: grammar, vocabulary, pronunciation, writing, listening and reading oriented for different levels of training. "Mondly" applies augmented reality in the form of virtual images as a teacher or a chat-bot for communication and understanding English. One of the features is that it identifies an individual's speech against the background of various noise and in such a way conditions for real life conversations with a virtual character are recreated. Student can use this application for better theory acquisition, pronunciation improvement, speech practice, memorizing new vocabulary and as a result mastering English in general.

"Rumii" is created as a platform for remote learning of a group and contains diversified tools: 3-D objects, a board, a screen demonstration. "Mind Map AR" facilitates learning a new topic in English, mastering new vocabulary by forming mental maps in augmented reality. This application makes it possible to connect associated notions using images, tags and represent these branches in a three-dimensional space. They are more effective than traditional lectures and make it possible to remember all the material easier and show abilities to creativity and independent work.

The BBC develops the application "Civilisations AR" that is devoted to study of artifacts of ancient civilisations, the period of Renaissance and others. It contains audio guides as well as texts in English. Thus, augmented reality opens up great opportunities in a foreign language teaching. It allows learners to know more about the masterpieces of the world culture and art and simultaneously improve their knowledge of English.

As for "ARki" and "Smart Reality", they will be informative for students specializing in architecture, design, construction because these applications allow to visualize architectural projects, represent three-dimensional models of buildings. ''Smart Reality'' can be used by combining interactive models with printed construction plans. Students follow different stages of construction, participate in this process, demonstrate each phase and as a result give a presentation of their projects using professional English vocabulary in their speech. Such work imitates a real professional activity where all the students of the group are involved in the process.

To clearly demonstrate the diversity of digital technologies using augmented reality in a foreign language teaching, we need to present the research data in the form of a table 1.

The results in the table 1 reveal that digital technologies are aimed at forming students' communicative competence that is considered one of the main purposes in teaching a foreign language. English teachers have a large arsenal of digital tools in their disposal. The choice of digital means depends on the direction of training in each particular group at the technical university.

For achieving the goal of the given research, it was decided to apply augmented reality technologies as well as traditional teaching methods in practical classes of a foreign language in one experimental group and compare the results—the level of formed skills in another group where augmented reality was not used. The pedagogical experiment for revealing high efficiency of augmented reality technology application was carried out in two groups of the Humanitarian Department in Rostov State Transport University. Both groups were estimated by the same indicators: a level of formed skills and communicative competence, an amount of allocated hours, motivation to learning English. The experimental group of students and the control one were practically identical. Students of the experimental group took part in approbation of efficiency theory of using augmented reality technologies in a foreign language teaching.

According to the discipline working program in terms of the curriculum of the training direction for the bachelor's degree 38.03.03 "HR", students of the first course should study the topic "The Target Language Country" in the second term. Within the framework of this topic, the English teacher gave a task for the control group using traditional teaching methods: read and translate the text, render it, make up a description of the country, its main sights, discuss this topic in pairs and so on. For another group—experimental—the task included besides working with a text and mastering new vocabulary—a project work with application of such augmented reality technologies as "Tour Builder" and "Tour Creator". All students of the experimental group were involved in work, their knowledge of English found practical application during creation of a tour around London. They chose different routes, added descriptions, found photos and video materials to them. The purpose of using modern technologies is not only to remember the new professionally oriented vocabulary or grammar forms of verbs, formulate a statement according to the rules, etc., that is to form the necessary skills and abilities in a foreign language, but to create real conditions for application of the received competences and to develop a professional competence.

The augmented reality technologies enable students to immerse in an English-speaking environment, find themselves in a target language country, get acquainted with famous places of interest firsthand. In addition to this, it is necessary to note that these means contribute to developing speech, listening skills in a foreign language, use of new vocabulary, as well as ability to group work, independent activity. The augmented reality technologies increase students' motivation in learning a foreign language, create positive emotional atmosphere in the lesson, promote communication in English, play an educational role of mutual understanding, support and help to each other.

As a result of application of such assessment methods, as a test, questioning, pedagogical observation, we came to the following conclusions. The students of the experimental group showed good results of formed skills, high motivation and interest in English. In comparison with the control group the achieved results in the experimental group turned out to be

Table 1. Using technologies of augmented reality in teaching a foreign language

The title of the technology	Formed Skills	Advantages	Disadvantages
Time XR Social AR&VR, a social platform	Listening and speaking skills	A teacher discusses different topics with students in a virtual world	Concentrated mainly on speech activity, other types of work are not represented
Altspace VR Community, a social platform	Listening and speaking skills	A teacher discusses different topics with students in a virtual world	Limited in use by speech activity for entertainment
Class VR, an educational platform	Reading, writing, listening, speaking skills	A teacher can plan and organize his own lesson or use the ready one, it is suitable for remote learning	The necessity of additional devices—glasses and a helmet
Engage (Immersive VR Education), an educational platform	Different skills	High school teachers have all tools for creating the content necessary for virtual classes, it is used for work in groups or individually in remote learning	Additional devices—glasses and a helmet, but their usage is optional
Virtual Speech, an educational course	Different skills, especially aimed at business communication	A teacher can combine traditional means of teaching and VR	It is not free
Jig Space, an application	Listening and speaking skills, skills of creative independent work	Students can create their own 3D presentations with the help of "Jig Space" or use different ready presentations of objects	Concentrated on creating presentation
Timeline JS, a service	Skills of creative independent work, including the usage of all abilities and deep knowledge of English	Students create their presentations, find and insert necessary information, video- and audio-materials from various sources	All information is presented in connection with dates, timeline

Table 1. (Continue)

The title of the technology	Formed Skills	Advantages	Disadvantages
Tour Builder, Tour Creator, services	Skills of creative independent work, including the usage of all abilities and deep knowledge of English	Students creating tours, excursions in English	Students must have an account in Google
Mondly, an application	Different skills	Students can communicate with a virtual character on different topics, an application identifies an individual's speech against the background of various noise	It can be used as an additional material for students' self-study, it can't identify voice perfectly
Rumii, an application	Listening and speaking skills	Teachers use a virtual space for remote learning	The necessity of additional devices—glasses and a helmet
Mind Map AR, an application	Lexical skills	It is used for creation mental maps	Focused on work on vocabulary
Civilisations AR, an application	Reading, writing, listening, speaking skills	Students learn of historical artifacts	Limited by historical topics

rather high. Therefore, the efficiency of augmented reality technologies in a foreign language teaching became evident and proved by practice.

4 Discussion

The conducted research has confirmed our hypothesis about the efficiency of using augmented reality technology in a foreign language teaching in high school. Nowadays we can observe the process of digitalization of all spheres of our life. The use of digital technologies in education is an essential part of our reality. Augmented reality is widely used in education of different levels from pre-school to high school and it has a positive impact on students. Its attractiveness finds an explanation in the main features that characterize augmented reality:

- interactivity;
- the effect of presence and immersion in the language environment;
- broadening the understanding of the surrounding processes and phenomena;
- sensory perception of updated data in a familiar environment.

Our study reveals availability of various digital means with augmented reality: social platforms, educational platforms, courses, services, applications—that can be used in practical classes of a foreign language. They are diversified and targeted at forming some definite skills, can be limited by a certain sphere of applying or specialization, but their main advantage is that they are distinguished by interaction with a virtual world in real time [13, 14]. This fact attracts students and promotes the effectiveness of educational process.

We agree with the ideas of Vat-U-Lan [15] concerning more effective assimilation of knowledge by students with using augmented reality technologies. These ideas find their confirmation in the process of the given research. The pedagogical experiment conducted in the course of the study has proved the high efficiency of augmented reality technologies. Students of the experimental group using besides traditional teaching methods digital means with augmented reality showed better results in formed skills, high motivation and interest in learning a foreign language.

We revealed that augmented reality in practical classes increases learners' motivation to study and self-study of a foreign language and greatly improves their knowledge. Therefore, students can apply their skills and abilities in practice with the help of augmented reality technologies. They take an active part in discussions, reading and listening practice and show creativity in their own presentations, creation of tours and other projects applying their knowledge of the English language, formed skills of research activities so essential in their future profession.

5 Conclusion

As a result of the given research, we should conclude that English teachers have a great variety of digital means with using augmented reality. Such modern means play a significant role in improving the educational process in a foreign language teaching. One of the most important functions of their usage is that they are targeted to forming all necessary competencies—speech, linguistic, communicative, sociolinguistic ones. Besides, the other function is connected with developing great motivation of students to learning English. These augmented reality technologies make them take an interest in a foreign language and be carried away with the process of learning grammar rules or some professional vocabulary that is rather complicated and even boring for students in traditional lessons, but quite exciting using augmented reality.

The conducted research has a great practical importance as it contains the detailed analysis of all augmented reality technologies acceptable for a foreign language teaching. They can be applied by educators during the classes. The further research activity is connected with the problem of reflecting augmented reality technologies in official documents with the purpose of their wide usage in high school.

References

[1] 21st century skills, Learning for Life in Our Times, URL `http://21stcenturyskills-book.com/resources/`

[2] D. Roopa, R. Prabha, K.J. Senthil, Materials today: Proceedings, **46(9)**, 3860 (2021), DOI: 10.1016/j.matpr.2021.02.294

[3] T. Joda, G. Gallucci, D. Wismeijerc, N. Zitzmann, Comput. Biol. Med., **108**, 93 (2019), DOI: 10.1016/j.compbiomed.2019.03.012

[4] T. Liao, P.F. Chang, S. YiLee, in *Technology and Health. Promoting Attitude and Behavior Change*, 109 (Academic Press, 2020), DOI: 10.1016/B978-0-12-816958-2.00006-X

[5] D.J. Thomas, Int. J. Surg., **36 A**, 25 (2016), DOI: 10.1016/j.ijsu.2016.10.003
[6] A. Klimovaa, A. Bilyatdinovaa, A. Karsakova, Procedia Comput. Sci., **136**, 5 (2018), DOI: 10.1016/j.procs.2018.08.232
[7] Ch. Samihah et al., Int. J. Hum. Comput. Stud., **134**, 44 (2019), DOI: 10.1016/j.ijhcs.2019.10.002
[8] H. Baiab, Q. Zhangab, Microprocess. Microsyst., 103421 (2020), DOI: 10.1016/j.micpro.2020.103421
[9] J. Garzónab, K. Silvia, B. Jaime, G.J. Pavónb, Educ. Res. Rev., **31**, 100334 (2020), DOI: 10.1016/j.edurev.2020.100334
[10] T.C. Hsu, Comput. Educ., **106**, 139 (2017), DOI: 10.1016/j.compedu.2016.12.007
[11] P. Coie and the XR Association, EdTech, URL https://edtechmagazine.com/k12/article/2019/08/survey-education-among-top
[12] R. Subedi, M. Shrestna, European Journal of Educational Technology, **3(1):1**, 1 (2020), URL https://www.researchgate.net/publication/339589094_Student_Friendly_Teaching_and_Learning_Environment_Experiences_from_Technical_Vocational_Educational_Training_Schools_in_Nepal
[13] J. Calvert, R. Abadia, Comput. Educ., **159**, 104005 (2020), DOI: 10.1016/j.compedu.2020.104005
[14] A.Y.M. Jonesa, E. Deanb, C. Hui-Chan, Comput. Educ., **54(4)**, 1193 (2011), DOI: 10.1016/j.compedu.2009.11.005
[15] Vate-U-Lan P., 2012 IEEE International Conference on Multimedia and Expo, 890 (2012), DOI: 10.1109/ICME.2012.79

Scientific and educational practice-oriented complex for training specialists in end-to-end technologies of space remote sensing of the Earth

Victor Soifer[1], *Vadim* Salmin[2], *Gennadiy* Anshakov[2], *Ivan* Tkachenko[2,*], *Sergey* Safronov[2], *Maksim* Ivanushkin[2], and *Anastasiia* Krestina[2]

[1]Department of Technical Cybernetics, Samara National Research University, 34, Moskovskoye shosse, Samara, 443086, Russia
[2]Department of Space Engineering, Samara National Research University, 34, Moskovskoye shosse, Samara, 443086, Russia

Abstract. This work is devoted to the development of a scientific and educational complex of competence and training of world-class specialists in end-to-end technologies of space remote sensing of the Earth. The complex is aimed at developing the competencies of students at all stages of the process of creating space-based means of remote sensing of the Earth—from design, production, testing and operation to processing space images and hyperspectral data and the formation of the final information product using modern digital technologies. End-to-end educational technology from the project to the final information product for a wide range of customers (data received from satellites are used by about 200 different Russian ministries and departments) formed the basis of the scientific and educational activities of Samara University.

1 Introduction

Space exploration of the Earth from space has been carried out for more than 60 years, providing people with an increasing amount of necessary and useful information. Without up-to-date and high-quality information about the state of objects and phenomena of interest related to the Earth's surface and the atmosphere, it becomes impossible to conduct effective activities in the field of meteorology, ecology, economics and many other types of human activities.

The development of Earth remote sensing (ERS) means is progressing at an increasing pace. In the period from 2006 to 2015, two to three dozen remote sensing satellites were put into orbit every year, and in the last five years there has been an exponential growth in the number of launched vehicles due to the deployment of multi-satellite space observation systems [1].

The most general trend is the increase in the frequency and resolution of the optical-electronic equipment of the ERS satellites, as well as the active development of new spectral ranges (ultraviolet and infrared parts of the spectrum). Obviously, the observation frequency is directly related to the swath of the spacecraft and, therefore, to the altitude of its orbit.

*e-mail: tkachenko.is@ssau.ru

Orbital altitude is not the only orbital parameter that determines the frequency (much depends on the inclination of the orbit and its eccentricity), but it is the need to achieve a high observation frequency that mainly determines the launch of satellites into higher orbits. The high frequency of observation is designed to identify rapidly occurring changes in objects of observation. High-resolution shooting in the "live video" mode is aimed at the same.

In recent years, an alternative way of ensuring a high observation frequency has been clearly identified based on the deployment of multi-satellite constellations of high-resolution optoelectronic observation in relatively low orbits [2]. One example of the implementation of this approach was demonstrated by the company "Planet Labs", which has formed a constellation of more than 200 spacecraft, mainly cubsats. They provide images with a resolution of 3–5 m of the entire surface of the Earth with a frequency of one day. At the same time, the trend of "large" satellites is aiming at the maximum resolution for space imagery of about 10 cm, and the trend of minisatellites generally tends to a resolution of 30–50 cm. An increase in spatial resolution and ground swath leads to a significant increase in the amount of information transmitted to ground receiving points.

Multispectral and hyperspectral observations made in the range of visible waves, ultraviolet and infrared ranges are the next natural step in the development of remote sensing. Collecting information in narrow ranges of the spectrum allows to achieve better object recognition based on their spectral characteristics. However, it should be understood that the amount of data produced by hyperspectral equipment increases many times over. This, of course, imposes new requirements on the characteristics of the transmitting and receiving path, on-board memory, and ground means information reception.

The above indicates significant changes in the approaches to the development of modern means of remote sensing [3]. The principle of consistency is still dominant and applies to all stages of the life cycle of spacecraft and monitoring systems: design (including ballistic)—manufacturing—testing—ensuring launch into a working orbit—target functioning—receiving and processing images—forming a thematic product. Thus, we are talking about the formation of an end-to-end technology for space remote sensing of the Earth. This requires new approaches in the training of specialists to ensure the development of advanced remote sensing means. Such specialists must possess a wide range of modern knowledge, skills and practical skills that allow the creation and effective operation of satellite and monitoring systems.

The main goal of this work is to provide training of world-class specialists with knowledge and competencies in all stages of the life cycle of space remote sensing systems. This makes it possible to solve the problems of staffing a scientific and technological breakthrough in the rocket and space industry.

2 Prerequisites for the creation of a scientific and educational practice-oriented complex

The advanced experience accumulated by Samara University in the training of highly qualified specialists is a catalyst for the introduction of innovative forms of scientific and educational activities at the university in the following areas:

- solution of fundamental and applied problems relevant to aerospace science, technology and engineering;
- transition to digital technologies for creating and supporting the life cycle of products (CALS-technologies) of aircraft engineering, engine building and space engineering;
- supercomputer and grid technologies for creating computer models of aircraft and space technology products, databases of electronic models of aircraft, engines and their systems;

- development of innovative production technologies: magnetic-pulse, electrochemical, ion-plasma, laser processing of materials, nanotechnology, technologies for rolling high-strength aluminum strips, vibration protection devices based on metal-rubber material, development of space biomedical devices;
- development of software and hardware for geoinformation systems, databases and services for high-precision, operational spatial modeling of territories based on Earth remote sensing data;
- personnel and scientific support for the creation of aerospace technologies, geographic information systems and Earth remote sensing systems;
- meeting the needs of the economy and national security with new geoinformation technologies and databases;
- integrated use of space images and Earth remote sensing data for global positioning systems;
- radical renewal and modernization of the content and forms of educational activities in the direction of intensifying the introduction of the results of scientific research into the educational process, the use of innovative educational approaches (polyprofessional, project, problem-search forms) that increase the motivation of students to acquire knowledge on their own.

Over the past years, Samara University has been consistently improving its performance in both Russian and international rankings. One of the most important results of the program for increasing the competitiveness of Samara University among the world's leading research and educational centers was a significant increase in the number of foreign students (from 150 people in 2012 to 1000 people in 2020), which ensures the export of Russian aerospace education abroad and promotes the achievements of Russian cosmonautics.

2.1 Cooperation with "Space Rocket Center "Progress"

The most important area of cooperation between the university and organizations in the rocket and space industry is the training of specialists and the creation of branches of base departments at enterprises. The result of this work was the long-term scientific and technical relations of the university with enterprises for targeted training and retraining of specialists, the involvement of leading specialists of the enterprise in giving lecture courses and supervising diploma design; training of scientific personnel through postgraduate study from among the specialists of enterprises.

Interaction between Samara University and the leading enterprise of the Samara rocket and space cluster Joint Stock Company "Space Rocket Center "Progress" (JSC "SRC "Progress") has a long history and is expressed in the following:

1. Targeted training of postgraduate and master's students for the "SRC "Progress".
2. Joint preparation and publication of papers, monographs [4].
3. Joint participation in large scientific and technical projects.
4. Participation of specialists of the enterprise in the work of branches of departments at the enterprise.
5. Joint organization and holding of scientific and practical conferences, symposia, seminars and educational events.

The enterprise has established a branch of the Department of Space Engineering. Classes are also held in the laboratories of strength, thermal vacuum, special tests of the research and testing complex of the enterprise. Students of the 4th and 5th years of the Department of Space Engineering have the opportunity to get acquainted with real technology at their workplaces. A unique innovative form of improving the system of university training of specialists is the conduct of industrial practice in Baikonur and Vostochny cosmodromes.

2.2 Cooperation with the Russian Academy of Sciences

Samara University maintains close cooperation with the Institute of Image Processing Systems (IIPSI) of the Russian Academy of Sciences (supervisor—President of Samara University, Academician of the Russian Academy of Sciences, Doctor of Technical Sciences, Professor V.A. Soifer). The Institute is integrated into the Samara University, actively cooperates with the university scientists in carrying out research work and successfully implements its achievements in the educational process—in the training of bachelors, masters, engineers and highly qualified personnel.

Samara University and IIPSI RAS jointly organized at the university and are successfully developing a laboratory of breakthrough technologies for remote sensing of the Earth. The main scientific areas of research at the RAS, in which scientists from the Samara University take part:

- computer optics, nanophotonics, optical information technologies and systems [5];
- image analysis and pattern recognition systems;
- geoinformation technologies.

The interaction of the university and academic structures, and primarily with the IIPSI RAS, plays a decisive role in improving the quality of fundamental training of highly qualified specialists, as well as the level of scientific research.

3 Scientific and educational practice-oriented complex for training specialists in end-to-end technologies of space remote sensing of the Earth

As part of the development of a scientific and educational practice-oriented complex of interdisciplinary study programs and laboratory testing facilities for training specialists in end-to-end technologies of space remote sensing of the Earth, interdisciplinary practice-oriented educational programs have been developed, tested and successfully implemented for 16 years. In addition, unique laboratory testing facilities have been created and integrated into the educational process, which allow for the development of the stages of end-to-end technology for creating remote sensing space assets.

3.1 Interdisciplinary study programs

The main programs developed with the participation of representatives of JSC "SRC "Progress" and IIPSI RAS, implementing the training of specialists in end-to-end remote sensing technologies:

- 24.05.01 "Design, production and operation of rockets and rocket-space complexes" (specializations: "Manned and automatic spacecraft and systems", "Modeling and information technologies for the design of rocket-space systems", "Rocket-space composite structures"), qualification—specialist;
- 24.03.01 "Rocket complexes and astronautics", qualification—bachelor;
- 11.05.01 "Radio-electronic systems and complexes" (specialization: "Onboard radio-electronic systems of rocket and space technology", "Radio-electronic systems of information transmission", "Antenna systems and devices"), qualification—specialist;

- 03.04.01 "Applied mathematics and physics" (specialization: "Mathematical modeling and information technology in photonics", "Space information systems and nanosatellites. Navigation and remote sensing of the Earth"), qualification—master;
- 01.04.03 "Applied Mathematics and Informatics" (specializations: "Mathematical Modeling and Computational Technologies", "Data Science", "Mathematical Methods for Modeling and Functional Design of Information Optical Systems and Devices", "Intelligent Data Analysis"), qualification—master.

Distinctive features of the developed educational programs are:

- enhanced fundamental training, which forms the basis for mastering special disciplines;
- high level of digitalization of both the educational process and the production process of creating remote sensing space vehicles;
- training in end-to-end technologies of space remote sensing, which forms knowledge and competencies throughout the life cycle of space technology products;
- the interdisciplinary nature of educational programs, suggesting a wide range of areas taught at the intersection of disciplines (flight dynamics and strength of spacecraft; design and optical systems; design and information technology);
- constant updating of the content of educational programs, taking into account the latest achievements of domestic and world space science and technology;
- the widespread use of individual educational trajectories, as well as the formation of conditions for the implementation of the "education throughout life" approach.

The developed programs provide personnel support by highly qualified specialists at all stages of the end-to-end technology for creating spacecraft for remote sensing of the Earth: from the design stage of a spacecraft to the stage of obtaining an information product.

3.2 Laboratory testing base

In the period from 2006 to 2021, unique laboratory testing facilities were created and integrated into the educational process, which allow the development of the stages of the end-to-end technology for creating remote sensing space assets, namely:

1. Center for receiving and processing space ERS information.
2. Supercomputer cluster "Sergey Korolev".
3. Manufacturing and testing complex of small satellite.
4. Ground control complex for small satellite of the "AIST" series.

Technical requirements for the purchased equipment for the installations being created were formed, methodological support was developed that allows using the above installations not only in the research and production process, but also deeply integrating them into the educational process, making them an integral part and the main tool for the acquisition of practical skills of working with modern equipment and technologies.

3.2.1 Center for receiving and processing space ERS information

The center is equipped with unique equipment: hardware and software complexes in the direct reception mode receive satellite imagery data from Terra/Aqua, SPOT 2/4, IRS-P5/P6,

EROS A/B, RADARSAT 1/2 satellites (figure 1). Since 2010, the center has become the first in Russia certified to work with RADARSAT-2 radar data.

Figure 1. Center for receiving and processing space ERS information

The center has developed an automated system for monitoring and assessing the economic potential of the agro-industrial complex of the Samara region. The work is entirely based on satellite images. An educational and research software package for space monitoring in the field of environmental monitoring, nature management and land control has been developed. The data of the center are used, for example, in a laboratory workshop, in course and diploma design, in dissertation research carried out by graduate students and doctoral students of the university.

3.2.2 Supercomputer cluster "Sergey Korolev"

The tasks of the cluster are scientific research and training of personnel using scientific and educational supercomputer and grid technologies, the creation of competitive models of new technology in conjunction with the enterprises of the aerospace cluster. The supercomputer is used to simulate products of rocket and space technology at all stages of the life cycle, to optimize technological, production processes, and enterprise management processes (figure 2). It is equipped with the most modern CAD/CAM/CAE/PDM/PLM systems and licenses for their academic and commercial use. The supercomputer is connected to the leading industrial enterprises of the aerospace cluster by dedicated fiber-optic lines.

Figure 2. Supercomputer cluster "Sergey Korolev"

The cluster's capacity is constantly growing, and the supercomputer currently provides a peak performance of 40 trillion floating point operations per second (40 teraflops).

3.2.3 Manufacturing and testing complex of small satellite

The manufacturing and testing complex (MTC) of small satellite was created in the period from 2013 to 2015 on the basis of the experimental research building of JSC "SRC "Progress", located on the campus of Samara University. The material equipment of the MTC makes it possible to carry out key stages of testing and experimental development, as well as assembly of small spacecraft (figures 3–6). In addition, the production and testing complex is a permanent site for laboratory, practical work, as well as practices. Students have the opportunity to work on the equipment used to create real samples of space technology, primarily small ERS spacecraft and nanosatellites.

Figure 3. Climatic chambers KTK-3000 (left) and Feutron 3524/58 (right)

Figure 4. Vibration Test System DataPhysics, LE-2016/DSA10-200K

Figure 5. High vacuum chamber

Figure 6. Coordinate Measuring Machine

3.2.4 Ground control complex for small satellite of the "AIST" series

The ground control complex (GCC) for small satellite was created in 2015 to ensure the operation of small satellite of the "AIST" series (figure 7) [6]. The staff of the center—students and postgraduate students—work in several main areas [7, 8]:

- reception, processing and in-depth analysis of telemetric information from the "AIST" series small satellite and the implementation of scientific research on its basis;
- development of projects for new small satellite based on unified platforms "AIST";
- implementation of the educational process and the development of new educational programs with deep implementation of the results of the functioning of the small satellite.

The implementation and modernization of individual disciplines and educational programs as a whole is carried out on an ongoing basis for the training of specialists, bachelors

https://doi.org/10.1051/shsconf/202213701026

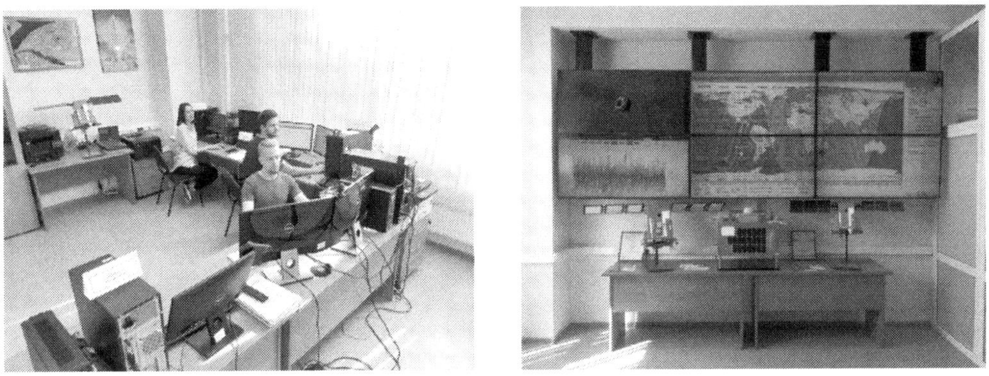

Figure 7. Ground control complex for small satellite of the AIST series

and masters of Samara University by introducing the results of the functioning of the constellation of "AIST" small satellites into the educational process.

3.3 Small satellite "AIST-2D"

The "AIST-2D" small satellite (figure 8) is the most striking example of the practical application of the developed scientific and educational practice-oriented complex of interdisciplinary educational programs and laboratory testing facilities for training specialists in end-to-end technologies of space remote sensing of the Earth. The design of the apparatus was carried out entirely in "digital form" using modern technologies of top-down design, which allows to ensure the simultaneous work with the product of different groups of performers and the constant "linking" of all parameters of the satellite [9]. The device was manufactured using modern technological solutions using high-tech production equipment. Ground experimental testing of the satellite was carried out in a short time and in the minimum volume, which at the same time ensured the required level of reliability.

Figure 8. The "AIST-2D: small satellite

The creation of the satellite by specialists from Samara University and JSC "SRC "Progress" (mostly graduates of Samara University) in a record-breaking time for the industry—32 months (from design to launch and obtaining information) is also a confirmation of the effectiveness of the developed methodological approaches to training. Such a speed of creation of fundamentally new ERS space assets is a world-class achievement.

The processing of images received from the apparatus is carried out using algorithms and software created by specialists from the Samara University and the IIPSI RAS in conjunction with specialists from the JSC "RSC "Progress" [10].

4 Results

The developed scientific and educational practice-oriented complex of interdisciplinary curricula and laboratory testing facilities is designed to train specialists in end-to-end technologies of a dynamically progressing industry—space remote sensing of the Earth.

Based on an interdisciplinary approach and the implementation of individual educational trajectories, the formation of students' competencies at all stages of the life cycle of the ERS space system is provided—from design and testing of a spacecraft to thematic image processing.

The end-to-end technology of creating ERS spacecrafts from the project to the final information product for a wide range of customers formed the basis of the scientific and educational activities of Samara University. This made it possible to provide training for specialists with knowledge, competencies and skills in all stages of the life cycle of the ERS space system, and to solve the problem of staffing a scientific and technological breakthrough in the rocket and space industry. The end-to-end educational training of specialists in space ERS is based on fundamentality, multidisciplinarity, the implementation of the concept of "education through research", the inextricable connection of education with industrial practice at the JSC "SRC" Progress" and the Baikonur and Vostochny cosmodromes, a high level of digitalization, the involvement of students in a practical solution tasks of creating spacecraft.

Since the realization of the project and the implementation of its results in the educational and pedagogical activities of Samara University (2006–2021), more than 3000 specialists have been trained, employed in the industry. The work was awarded the Yuri Gagarin Russian Government Prize in the field of space activities in 2021.

References

[1] J. Radtke, C. Kebschull, E. Stoll, Acta Astronautica, **131**, 55 (2017)

[2] V. Salmin, V. Kurenkov, S. Safronov, I. Tkachenko, A. Yakischik, M. Ivanushkin, S. Volgin, A. Krestina, J. Phys.: Conf. Ser., **1745**, 1 (2021)

[3] A. Kirilin, R. Akhmetov, S. Tkachenko, N. Stratilatov, V. Abrashkin, S. Safronov, Y. Zheleznov, E. Shakhmatov, V. Soifer, A. Prokofev, V. Salmin, N. Semkin, I. Tkachenko, Procedia Engineering, **104**, 76 (2015)

[4] G.P. Anshakov, V.V. Salmin, K.V. Peresypkin, A.S. Chetverikov, I.S. Tkachenko, J. Phys.: Conf. Ser., **1096**, 1 (2018)

[5] V.A. Soifer, D.L. Golovashkin, L.L. Doskolovich, *Computer Design of Diffractive Optics* (Woodhead Publishing Limited, Cambridge, United Kingdom, 2013)

[6] A. Kirilin, R. Akhmetov, S. Tkachenko, N. Stratilatov, V. Abrashkin, S. Safronov, Y. Zheleznov, E. Shakhmatov, V. Soifer, A. Prokofev, V. Salmin, N. Semkin, I. Tkachenko, Procedia Engineering, **104**, 43 (2015)

[7] M.A. Ivanushkin, I.S. Tkachenko, S.L. Safronov, I.V. Kaurov, S.S. Volgin, J. Phys.: Conf. Ser., **1368**, 4 (2019)

[8] S.I. Tkachenko, V.V. Salmin, I.S. Tkachenko, S.L. Safronov, I.V. Kaurov, M.D. Korovin, M.A. Ivanushkin, S.S. Volgin, CEUR Workshop Proc., **1638**, 691 (2016)

[9] A.N. Kirilin, R.N. Akhmetov, E.V. Shakhmatov, S.I. Tkachenko, A.I. Baklanov, V.V. Salmin, N.D. Semkin, I.S. Tkachenko, O.V. Goryachkin, *Experimental and technological small spacecraft "AIST-2D"* (SamSC RAS, Samara, 2017)

[10] V.A. Soifer (eds.), *Computer Image Processing, Part II: Methods and algorithms* (VDM Verlag Dr. Müller, Saarbrücken, 2009)

On the problem of motivation of technical university students in the case of group implementation of an interdisciplinary term project

Elena Vlasova[1,*], *Vladimir* Popov[1], and *Oleg* Pugachev[1]

[1]Bauman Moscow State Technical University, 5/1, 2nd Baumanskaya st., Moscow, 105005, Russia

Abstract. The authors consider the problems concerned with the mathematical constituent of implementation of interdisciplinary semester projects by groups of students. Significance of using the method of projects for creation of the innovative environment in a technical university, as well as for formation of professional competences of future engineers is demonstrated. The authors have analyzed the techniques of student groups' work on term projects and reflected the role of a teacher in the process of implementation of such projects. It is stressed that one of the major components of technical university education is fundamental training of engineering students based on deeper and wider cycle of mathematical and common engineering disciplines. It provides close interdisciplinary relations between the mathematics course and engineering disciplines, increases the motivation to master mathematics, helps the students to recognize the need to gain profound mathematical knowledge, and forms the skills of creative application of the gained knowledge to successful mastering their engineering professions chosen. It is demonstrated that, in a technical university, an interdisciplinary term project is in essence a task that unites various scientific and technical disciplines and demands engineering and technical knowledge as well as mathematical methods of task solution: applied methods of mathematical modeling, methods of probabilistic and statistical analysis of data, and various types of mathematical packages.

1 Introduction

Training of specialists with a higher technical education necessitates development of students' creative skills, stepwise promotion of each student from his/her personal starting level of development to the certain necessary level of qualification, that is obligatory for all graduates. Creativity is an independent work; therefore, it is necessary to pay close attention to students' independent work and to supervise their work. However, any supervision is concerned with imposing certain restrictions on the supervised person's activities. This necessitates regulation of education and creativity, but the regulation, in its turn, must not impede development of the person's creative skills. Thereby, we face the challenge to optimize the supervision of the students' independent work in the process of development of their creative skills; elaboration of the qualification requirements for graduates of universities, and for graduates of technical universities, in particular. A young engineering specialist must be

*e-mail: vlasovaea@bmstu.ru

able to organize an effective work of the production team under his/her management, must have skills of organization and innovative development, must be familiar with the methods of scientific and practical organization of labor. Therefore, a specialist with a diploma of mechanical engineering must be independently able to construct certain parts of machines, to assign an economically viable technological processing, to perform necessary calculation of strength, to plan and perform a technical experiment, and to make a report.

Every engineer needs the aforementioned skills since the first days of his/her work at factories, design bureaus, technological departments, workshops, or laboratories. These skills are also necessary for performing scientific research, design and producing new articles, and technological processes. The special characteristic of all these skills is creativity. Therefore, one of the general purposes of education, and of technical education in particular, is development of cognitive and creative skills of students. Learning from creative experience is possible only via practical activities of the person. Hence a student's personal activities on the training material and his/her individual independent work or work in a group play the crucial role in development of his/her creative skills.

One of the most important components of technical university education is fundamental training of engineering students based on deeper and wider cycle of mathematical and common engineering disciplines. The mathematical curriculum in a technical university is an example of harmonious combination of pure and applied mathematics. It provides close interdisciplinary relations between the mathematics course and engineering disciplines, increases the motivation to master mathematics, helps the students to recognize the need to gain profound mathematical knowledge, and forms the skills of creative application of the gained knowledge to successful mastering their engineering professions chosen.

Since junior years, it is necessary to motivate students to gain mathematical knowledge and to associate it with future professional activities, stressing its practical significance. For this purpose, it is necessary to actively apply *term projects* in mathematical disciplines built into consolidated interdisciplinary *individual* or *group projects* [1–3]. The subject of such a project, especially of a group project, ought to be directed to stimulate cognitive interest during education, in order to make the student conscious of the need of such activity and its practical feasibility. An unfortunately chosen subject of a term project (with results which students do not know in advance in general terms) often leads students to implement it formally and blankly and does not arise their considerable cognitive interest.

During implementation of group projects, students must learn to analyze and systematize materials of the subject, to construct the logical reasoning leading to the result, to formulate purposes and conclusions adequate to the task posed. All this leads to formation and development of competences, such as determining the status of the problem posed, independent elaboration of the statement of work, planning the stages of its implementation, and forecasting its results.

2 Technique of group exercises

Working together in a group is one of the forms of education process. Such form of work is often applicable during implementation of various term projects [4, 5]. Uniting persons into groups is a natural phenomenon.

In recent years, the forms of working in groups are quite widespread, so that there is no doubt about their attractiveness and usefulness. However, students working together in a group are not always optimistic about such kind of unification, due to the following reasons: many of the students have no experience of working in groups; they may have negative experience; there may be opinion that discussions in groups are useless; individual problems concerned with activity in a group, etc.

It is clear that working in groups has both positive and negative aspects. As benefits of working in groups, one can regard e.g. that unification is purposed to achieve certain common goals that are more difficult and expensive to achieve individually. In a group, a person satisfies several needs instead of one, performs several tasks instead of one. In the other hand, e.g. it is known that in groups, it is comfortable for a weak student and it is hard for a strong one. The former may use the group's resource to compensate his/her weakness; the latter has to waste energy for other members of the group.

In a group, the quantity of communication acts increases. Often it makes the process of group decision making longer and more difficult. The so-called effects of "groupthink" sometimes arise in groups: the group members feel total confidence that the group is right and its opinion is unshakable, and absence of doubts and interest to alternative opinions. In such situation, he group disengages itself from development. In unbalanced groups, some conflicts may arise and become obstacles to implementation of the tasks. The students' inner tension may arise and cause irritation towards the teacher or against the tasks given by the teacher.

Along with other factors, effectiveness of working of the group depends on its size, and this dependence is not uniform. The only is inevitable—a large group splits into several smaller ones.

Another peculiarity lies in the fact that sometimes activity of those who were working better decreases during working in groups. In one hand, the spirit of collectivism and competition makes the members of the group to increase productivity of their labor. In the other hand, there is a tendency to work in the group slightly less than in the case of individual work.

There is no unanimity of opinions concerning the optimal quantity of group members. Any group is considered to include at least four persons. In this case, one can organize operational work in the group. However, a group of five to seven students is optimal. Groups of such size are the most effective when solving *interdisciplinary tasks* of term projects. At the same time, such group is small enough in order to recognize the contribution of each student. If the number of group members is odd, it is easier to make group solutions.

The teacher stands in the center of group formation and its character. The most significant is the teacher's role in formulation of *group aspirations*, i.e. stating the problem and the purpose, and control of their implementation. The teacher must motivate the students to achieve the purposes stated.

The teacher's significant role consists in creation of the *group mood*, which is to facilitate joint efforts and success of the group. It should be noted that typically group moods are more dynamic than individual ones, hence the former are more convenient to manage.

Elaboration of the *group opinion* is another important constituent part of the teacher's work. The group opinion is the integrated opinion shared by all the group members; it is generalization and unification of individual opinions during the work on the project. The teacher ought to pay attention to the students' statements and remarks and to monitor the process of formation of the group opinion closely.

List the complex of measures that can provide increase of education outcomes of group implementation of term projects.

1. Elaboration of the "bank" of term project tasks of applied nature, taking the specific of professional interests of future engineers in account.
2. Specification of purposes of implemented term projects from the viewpoint of qualification requirements to graduates of the certain university, and delivering the message of these purposes to the students.
3. Supporting systematic independent students' work on the project and their sufficient training for its implementation.

4. Alignment of the students' activities at implementation of the term project, and optimization of their working time.
5. Increasing the teacher's productivity.
6. Increasing educational influence onto the students, and systematical parenting skills to them, in order to provide successful work on the project.
7. Monitoring of the educational effect of the project implementation.
8. Organization of group activities and interaction between its members.
9. Using innovative methods and technologies during implementation of term projects, in order to make the students' work more active and systematic and to increase their motivation to achieve results of practical significance.
10. Evaluating each student's contribution in the group outcome.
11. Accountability of the outcomes obtained.

3 Mathematical constituent of a group interdisciplinary term project

A group term project is a means of goal achievement via elaboration of the problem and obtaining a concrete practical and applicable result by students working in a small group.

In a technical university, an interdisciplinary term project is in essence the task that integrates various scientific and technical disciplines and demands engineering and technical knowledge as well as mathematical methods of task solution: applied methods of mathematical modeling, methods of probabilistic and statistical analysis of data, and various types of mathematical packages [6].

An important constituent of such a project is its subject. The authors have elaborated a large "bank" of such tasks having applicable nature [7–10] and concerned with professional specific of the university graduates. As the research, they have chosen air and space technical units, e.g. various types of motors, turbines, and flying machines.

Interdisciplinary term projects, due to their large volume, from the very beginning, are to decompose into project modules, and the latter are to split into several submodules to practice in small groups. The members of each project group are constantly interacting and hence, there is interaction among members of the small groups.

Implementation of term projects in groups takes place under supervision of teachers belonging to both mathematical and specialized departments, quite independently. In time determined by the agreed plan, they deliver information about the work accomplished. Quantity of students in the groups is determined based on the technical task of the term project.

Usage of innovative methods and techniques [4, 11] and various types of mathematical packages in the process of implementation of term projects allows enhancing and systematizing the students' work, to increase their motivation to achieve outcomes of practical significance.

Another serious motivating stimulus is a student's individual (*cumulative*) rating introduced by the authors [12]. The rating aims at differentiation of evaluation of results of the student's work. The cumulative rating plays a special role in the situation where students implement a term project in a small group, and it is important to evaluate each performer's contribution into the outcome.

Implementation of a term project starts from drawing up a detailed plan of solution of the problem posed; hereby it is necessary to involve the students of the group. Such kind of plan must contain the stages of implementation of the project; moreover, from the teacher's viewpoint, it must be able to realize the methodical objectives of the students' group work, i.e., use of theoretical knowledge, abilities and skills achieved. From the experience of the authors' elaboration of such projects (with mathematical constituent including probabilistic and statistic methods) the plan may include the following items.

1. *Construction of the mathematical model.* Students must be independently able to compose the equations describing the physical phenomena of the problem, to compare the dependencies in order to establish qualitative correctness of the chosen model. Herewith the students obtain the skills of independent search and systematization of information, of analysis and choice of integral-differential equations describing the mathematical model of the problem, and the capability to use interdisciplinary connections between mathematics and mechanics.

2. *Construction of the estimate using maximum likelihood.* It is required to compose the likelihood function and to use it in order to calculate the values of the estimates.

In this item of the plan, the students use the method of maximum likelihood and the principle of invariance, the skills of analytical and numerical calculation of estimates of parameters of distributions.

3. *Comparison of the estimates obtained.* It is necessary to make qualitative and quantitative comparison of the results obtained and to explain the connections discovered.

Herewith the students obtain the skills of systematic analysis of the phenomena under consideration, of formulation and argumentation of the patterns identified, to pay attention to particular phenomena of the results obtained.

4. *Constructing the confidence interval for the model parameters.* Students ought to be capable of using the well-known formulas of regression analysis and the property of Gaussian distribution. In the performance of this task, students obtain the skills of independent search and systematization of information, using formulas of confidence intervals in the Gaussian model.

5. *Statistical analysis of the regression model.* It is offered to students to check the hypotheses of validity of the model (using Fischer's test) and of significance of each individual coefficient by means of the confidence intervals constructed.

In the performance of this item, the students use their knowledge of theory of statistical criteria and the skill of checking statistical hypotheses by means of criteria they know.

6. *Estimating the correlation coefficient.* It is required to calculate the estimate for the correlation coefficient between the biases and to construct its confidence interval; draw conclusion from the results of the research; calculate values of the residuals and create the histograms with various numbers of grouping intervals chosen. Herewith the students acquire the skills of search and usage of known analytical solutions, analysis of the obtained results and drawing conclusions based on them, and checking statistical hypotheses by means of the confidence interval constructed.

7. *Account on the mathematical constituent of the project.* It is necessary to draw up the report with formulation of the problem posed, mathematical methods of its solution and the conclusions.

4 Organizing laboratory work in a small group

A great importance for successful technical creativity of an engineer is capability of organizing and carrying out experiments. When studying a range of engineering courses, students carry out laboratory works (workshops) purposing to reinforce and advance their theoretical knowledge and to get acquainted with the equipment and its handling. However, often the workshop does little to obtain the goals mentioned: the students are not trained enough in theory and have no skills of handling the equipment; they are not able to process the results of their experiments and prepare the reports, their work is devoid of proper systematic.

Often students' attitude to the workshops is formal and blank. One of reasons of such attitude is unfortunate task subject, which does not encourage creative approach and interest in the outcome. The situation compounds by the fact that the students' attention does not usually direct to the search for the causes of outliers of values of the parameters they found from the values given in tables in handbooks. Without understanding that the values of physical constants given in the tables are results of statistical processing of results of huge series of thoroughly implemented measurements, some students try to modify results of an individual experiment "to make it fit the correct answer". This causes direct harm to training of future specialists by forming their distorted view of the purposes, methods and possibilities of laboratory studies.

In fact, the educational result obtained by the workshop is not clear, it does not undergo analysis, but only the fact of defense of the report of the work accomplished is marked. Later, as a rule, the written reports received are not used anywhere, instead they are being destroyed. All this produces students' formal and blank attitude to the important practical part of the educational process, which requires much energy and time expenses.

These facts imply that it is necessary not only to improve subjects of the workshops, but also to increase educational effectiveness of the students' work. To the same extent, it is true with respect to term projects of future engineers.

The complex of measures directed to the following can provide increase of educational outcomes of the term projects and workshops of technical university students.

1. Practical concretization of purposes of the works from the viewpoint of their subject and qualification requirements for graduates of a certain university.
2. The students' theoretical preparedness for accomplishing the certain works.
3. Strengthening educational influence on the students, systematic promotion of their skills, helping their successful work.
4. Providing further practical usage of the results obtained by the students.
5. Control of educational effect of accomplishing he work.

Consider an example of organizing of the workshop "Numerical solution of boundary value problems for one-dimensional equation of heat conduction" for the third year students of Bauman Moscow State Technical University (BMSTU) [13], training on the specialty "Applied mathematics". The purpose of the work is to study the integral-interpolation method of constructing difference schemes for solution of boundary value problems for one-dimensional equation of heat conduction. Performance of the work is suggested to a small group of three to five students.

On the first stage, the students have to choose a one-dimensional mathematical model describing heat conduction in a thin rod of fixed length with thermally insulated side surface, with given thermal regime on its ends. Next, they have to compose the differential equation of parabolic type describing the process of heat conduction with several variants of boundary conditions. For solution of the stated problem, it is required to construct difference schemes using the integral-interpolation method, and to write a program or numerical solution of each problem. Herewith, they consider the two cases of dependences of coefficients of heat conducting: on the spatial co-ordinate and on the temperature. On implementation of the work, it is required to find the point in time, when the temperature in the middle of the rod would be the largest. The calculation is to perform with the required degree of accuracy. The calculation results are to be illustrated with animation or a series of graphs, to make it possible to observe evolution of heat distribution in the rod.

The report shall contain the following items.

1. Stating the problem.
2. Description of the mathematical model, justification of its choice.

3. The difference scheme.
4. Description of the solution method.
5. Test examples demonstrating operating capability of the program.
6. Analysis of the difference scheme under consideration, including research of its monotonicity.
7. Research of stability of the method of solution of the system of difference equations.
8. Analysis of the results, comparison of solutions obtained by the direct scheme, the indirect scheme and the symmetric scheme, justification of the choice of a spatial step and a time step, comparison of effectiveness of the schemes under consideration.
9. Presentation of the results by graphs and animation.
10. Description of the group's and each student's contribution in accomplishing the laboratory work.

For each laboratory work, the teachers, for their part, must provide methodological guidelines containing the necessary general theoretical material, practical recommendations concerning the choice of a certain method for solution of specific tasks, the list of control questions to prepare the students' work defense, examples of allocation of tasks and duties in the group. The teachers must organize the defense of the laboratory work so that each student in the group would have to demonstrate knowledge of the theory of numerical methods to the extent necessary and the skill of applying this theory in practice.

5 Discussion and conclusion

Group implementation of an interdisciplinary term project should be preceded by preparing the complex of teaching methods and materials, educational guidelines, calculation patterns etc. When preparing the materials, it is important to use modern information and communication technologies, and various types of mathematical packages necessary for students in their future engineering activity.

The mathematical constituent of a group interdisciplinary term project in a technical university involves a serious fundamental mathematical training of future engineers, in order to provide integration with engineering disciplines. It is necessary to increase the students' motivation to mastering and practical application of mathematical knowledge in engineering calculation, for successful mastering their future profession.

Active application of *term projects* in mathematical disciplines built into consolidated interdisciplinary *individual* or *group projects* is necessary. Hereby, subjects of these projects should purpose to stimulate cognitive interest during the education, in order to make a student conscious of the need of such activity and its practicality.

As the result of accomplishing these tasks, the students are to learn the following.

1. Analysis and systematization of thematic materials.
2. Independent choice from alternative methods of solution of the problem.
3. Mathematical modeling of objects and processes according to templates, including usage of standard application packages.
4. Applying the suitable mathematical apparatus for solution of the problems stated.
5. Composing reports on the subject of the research accomplished.

All these lead to formation and development of the following competences: identification of the state of the problem, independent elaboration of the technical task, planning the stages of its implementation, and forecasting its results.

References

[1] Y.E. Krasnov, in Method of projects: Scientific and methodological collection. Series: Modern technologies of University Education, **2**, 197 (National Institute for Higher Education, Belorussian State University, 2003), URL `http://elib.bsu.by/handle/123456789/11590`

[2] H.-Y. Wang, I. Huang, G.-J. Hwang, Journal of Computers in Education, **3(1)**, 33 (2016), DOI: 10.1007/s40692-015-0047-9

[3] S. Han, R. Capraro, M.M. Capraro, Int. J. Sci. Math. Educ., **13(5)**, 1089 (2015), DOI: 10.1007/s10763-014-9526-0

[4] E. Vlasova, V. Popov, Bulletin of Moscow Region State University. Series: Physics and Mathematics, **1**, 101 (2017), DOI: 10.18384/2310-7251-2017-1-101-112

[5] E. Penkovskikh, Educational Studies Moscow, **4**, 307 (2010), DOI: 10.17323/1814-9545-2010-4-307-318

[6] E. Vlasova, N. Mezhennaya, V. Popov, O. Pugachev, Bulletin of Moscow Region State University. Series: Physics and Mathematics, **4**, 114 (2017), DOI: 10.18384/2310-7251-2017-4-114-128

[7] E. Vlasova, V. Zarubin, G. Kuvyrkin, *Approximate methods of mathematical physics* (BaumanPress, Moscow, 2004)

[8] V. Zarubin, G. Kuvyrkin, I. Stankevich, *Mathematical models of applied mechanics* (BaumanPress, Moscow, 2016)

[9] E. Vlasova, V. Zarubin, G. Kuvyrkin, *Mathematical models of processes of heat conductivity* (BaumanPress, Moscow, 2016)

[10] I. Savelyeva, I. Stankevich, *Mathematical modelling of processes of heat conductivity by the finite element method* (BaumanPress, Moscow, 2018)

[11] in Innovative technologies in science and education: Proceedings of the III Intern. scient.-pract. conf., 23 October 2015, Cheboksary, Russia, **3(3)** (2015)

[12] E. Vlasova, V. Popov, Bulletin of Moscow Region State University. Series: Physics and Mathematics, **4**, 86 (2019), DOI: 10.18384/2310-7219-2019-4-86-94

[13] I. Marchevskiy, O. Shcheritsa, *Numerical methods of solution of problems of mathematical physics* (BaumanPress, Moscow, 2016)

Professional and public accreditation for professional educational programs in the field of rocket engineering and space activities in Omsk State Technical University

Aleksey Yakovlev[1,*] and *Valentin* Belkov[2]

[1]Aircraft and Rocket Building Department, Omsk State Technical University, 11, Mira prospekt, Omsk, 644050, Russia

[2]Mechanical Engineering Department, Omsk State Technical University, 11, Mira prospekt, Omsk, 644050, Russia

Abstract. The article is devoted to the professional and public accreditation in Omsk State Technical University for bachelor's and master's degree programs "Rocket Systems and Astronautics". A brief description of the university is given. The objectives of accreditation are stated. The procedure for its implementation is described. The evaluation criteria and the results of the accreditation are given. The main comments identified by experts and the strengths of educational programs are presented.

1 Introduction

Professional and public accreditation (PPA) of the main professional educational programs, basic and (or) additional professional training programs is a recognition of the training quality and level of graduates who have completed such educational programs in a specific organization engaged in educational activity meeting the requirements of professional standards, labor market requirements for specialists, workers and employees of the relevant profile [1].

Professional and public accreditation is an effective tool for regulating and improving the quality of education, recognition of the high level of specialists training by the professional community. Passing an independent accreditation allows an educational organization to declare the quality of training specialists, thereby increasing its competitiveness in the market of educational services, as well as to ensure the employment of its graduates [2].

At the beginning of 2019, Omsk State Technical University made a decision on the need for independent accreditation for the bachelor's and master's degree programs of an integrated group of specialties and fields 24.00.00 "Aviation and Rocket and Space Technology". The objectives of the PPA were:

- to demonstrate commitment to the high quality of educational services and training specialists with higher education, public statement on the high quality level of specialists training;
- to obtain an independent qualified quality assessment of educational programs and specialists training;

*e-mail: yakovlev@omgtu.ru

- to analyse the strengths and weaknesses of the implemented educational programs, personal recommendations on updating educational programs from the State Corporation "ROSCOSMOS", one of the largest employers' associations;
- to ensure and improve the employment of graduates.

At that time, OmSTU had only two educational programs accredited by the Association for Engineering Education of Russia and the European Network for Accreditation of Engineering Education (ENAEE). It was the master's program 28.04.02 "Nanoengineering" and the bachelor's program 18.03.01 "Chemical Technology". There was no experience of public accreditation for bachelor's and master's degree programs of related specializations at the university at that time.

OmSTU application for professional and public accreditation of educational programs in the areas of training in rocket technology and space activity 24.03.01 "Rocket Systems and Astronautics" (profile "Rocket Engineering") and 24.04.01 "Rocket Systems and Astronautics" (profile "Rocket Design and Construction") was submitted to the expert organization—Agency for Professional and Public Accreditation and independent assessment of qualifications (LLC "Profaccredagentstvo", Moscow).

At the stage of considering the application by LLC "Profaccredagency", a positive decision was made to conduct PPA based on the fulfillment of the following conditions:

- the declared educational programs correspond to the type (types) of professional activity for conducting PPA in the Council for Professional Qualifications in the field of rocket technology and space activity;
- the availability of the stated educational programs in the license for the implementation of OmSTU educational activity;
- the preparation for the declared educational programs is carried out by the applicant for a period not less than the one established for the development of the educational program (there is at least one issue).

After the decision was made to conduct the PPA, the expert organization formed a schedule for an accreditation examination and gave recommendations for self-examination and preparing a package of materials for educational programs. The preparation of a report on programs and materials self-examination for accreditation examination [3] was carried out for two months by the OmSTU department "Aircraft and Rocket Building", which provides training in the areas declared for accreditation.

A group of independent experts was formed to conduct the accreditation examination by LLC "Profaccredagency", the composition of which was approved by the Accreditation Council:

1) Filimonov Alexey Sergeevich, an independent expert, Associate Professor of the Department SM-12 "Technologies of Rocket and Space Engineering", the Moscow State Technical University named after N.E. Bauman (National Research University), Candidate of Technical Sciences;

2) Mishchenko Vladimir Ivanovich, an independent expert, Head of the Department of Geometric Information and Cold Pressing No. 329 at IG "Polyot"—the branch of the Khrunichev State Research and Production Space Center in Omsk;

3) Yakovleva Svetlana Nikolaevna, a project supervisor from the accrediting organization, the chief specialist of the All-Russian Industrial Association of Employers "Union of Employers of the Rocket and Space Industry of Russia";

4) Baranova Natalia Vasilievna, a project manager, Director of the expert organization "Profaccredagency", Candidate of Pedagogical Sciences.

During the period of the accreditation examination (March–April 2019), the group of independent experts studied the materials submitted by OmSTU, formed a list of comments and clarifications. On April 16–17, 2019, an in-person visit of experts (on-site accreditation examination) took place, during which the expert group conducted interviews with the heads of the university, heads of educational programs, teachers of the Department of Aviation and Rocket Engineering, employees of the quality management system support group and the Regional Center for Graduate Employment Assistance, students, key partner employers, graduates, got acquainted with the university material and technical base and documents.

The educational programs presented for the PPA were developed in accordance with the Federal State Educational Standard:

- the field of education 24.03.01 "Rocket Systems and Astronautics" (bachelor's degree level), approved by the Ministry of Education and Science of the Russian Federation Order No. 1430 (dated December 4, 2015) and updated taking into account professional standards (PS) 25.028 "Process Engineer for Assembly Operation in Aerospace Industry", approved by the Ministry of Labor and Social Protection of the Russian Federation Order No. 997n (dated December 3, 2015) (registered with the Ministry of Justice of the Russian Federation on December 31, 2015, No. 40485) and PS 25.045 "Structural Engineer in Rocket Science", approved by the Ministry of Labor and Social Protection of the Russian Federation Order No. 939n (dated December 2, 2015) (registered with the Ministry of Justice of the Russian Federation on December 31, 2015 No. 40419);
- the field of training 24.04.01 "Rocket Systems and Astronautics" (master's degree level), approved by the Ministry of Education and Science of the Russian Federation Order No. 164 (dated March 6, 2015) and updated taking into account PS 25.045 "Structural Engineer in Rocket Science", approved by the Ministry of Labor and Social Protection of the Russian Federation Order No. 939n (dated December 2, 2015) (registered with the Ministry of Justice of the Russian Federation on December 31, 2015 No. 40419).

The heads of the main educational programs are:

- 24.03.01—Yakovlev Alexey Borisovich, Head of the OmSTU department "Aircraft and Rocket Building", Candidate of Technical Sciences, Associate Professor, Academician of the International Academy of Refrigeration;
- 24.04.01—Trushlyakov Valery Ivanovich, Professor at the OmSTU department "Aircraft and Rocket Building'', Doctor of Technical Sciences, Adviser to the Russian Academy of Rocket and Artillery Sciences, Director of the scientific and educational center ''Space Ecology''.

The number of students enrolled in the program 24.03.01 is 77 students, all at the expense of state budget, including students studying under employer-sponsored financing agreements (38 students).

The number of students enrolled in the program 24.04.01 is 25 undergraduates, all at the expense of state budget, including those studying under employer-sponsored financing agreements (6 students).

2 General characteristics of the educational organization

Omsk State Technical University (OmSTU) traces its history back to 1942. The university was organized as the Omsk Machine-Building one on November 16, 1942 by the Decree No. 1828 of the Council of People's Commissars of the USSR on the basis of the evacuated Voroshilovgrad Evening Machine-Building Institute. In 1963, by Order of the Ministry of Higher and Secondary Education of the RSFSR, the Omsk Machine-Building Institute was reorganized into Omsk Polytechnic Institute.

In 1965, the first recruitment for the specialty "Mechanical Equipment for Automated Installations" was carried out. In 1967 the department "Automated Installations" was established to train engineers for rocket launching and system performance. In 1985, the range of rocket and space specialties was expanded by the ones "Rocket engines" and "Rocket Building". In 1988, in order to train personnel for Omsk enterprises in the rocket and space industry, the departments "Flight Engines" and "Flight Building" were organized.

On July 5, 1993, by Order of the State Committee of the Russian Federation for Higher Education, Omsk Polytechnic Institute was renamed Omsk State Technical University.

In 1996, two dissertation councils were opened at the university for the defense of candidate dissertations in scientific specialties related to aviation and rocket and space technology. Since 2004, OmSTU has been training bachelors, and since 2008 masters in the areas of aviation and rocket building. In 2004, the departments of the rocket and space field were combined into a single department "Aviation and Rocket Building", which currently trains about 400 engineers, bachelors and masters for the needs of the aviation and space industry of Russia.

Since 2008, OmSTU has adopted a comprehensive development program by creating resource centers. OmSTU has set a course for an innovative-research type university with a developed scientific potential that trains in-demand and competitive specialists. The achieved results were the foundation and starting point for the "Strategic Development Program for 2012–2016", with which the university became the winner in the Russian competition of development programs among 248 universities and became one of the 100 best universities in Russia. In March 2016 OmSTU received the status of the flagship university by joining the Omsk University of Design and Technology. In April 2018, OmSTU successfully passed the state accreditation for educational activity in the main professional educational programs.

At the period, the university implements 33 integrated groups of training areas. There are 183 main professional educational programs of higher education, of which: 73 are bachelor's degree programs, 46 are master's degree programs, 12 programs for specialists and 52 educational programs for graduate students.

The main competitors of OmSTU in the Omsk region are: The Siberian State Automobile and Highway University; Dostoevsky Omsk State University; Omsk State Transport University; Omsk State Agrarian University named after P.A. Stolypin.

In the Siberian and Ural regions, according to the IGSA educational programs 24.00.00 "Aviation and Rocket and Space Technology", the main competitors are Novosibirsk State Technical University; Biysk Technological Institute; Reshetnev Siberian State University of Science and Technology (Krasnoyarsk); Perm National Research Polytechnic University; South Ural State University (Chelyabinsk).

The main influx of students to OmSTU is from the city of Omsk and Omsk region, the Republic of Kazakhstan, Novosibirsk Region, Tyumen Region and Khanty-Mansi Autonomous Okrug, Tomsk and Kurgan regions.

Today Omsk State Technical University is a modern technical university with a developed educational, scientific and industrial infrastructure, the largest university in the Omsk region. OmSTU is the driver of implementing the cluster model for industrial development of the Omsk region, the flagship regional university, distinguished by [4]:

- the use of modern domestic and foreign scientific achievements in the educational process;
- the orientation to close scientific and educational contacts with engineering companies and manufacturers of modern technology;
- the innovative infrastructure: research and educational resource centers, research laboratories, student design bureaus;
- a powerful human capital: highly qualified teaching staff and experienced researchers; support program for young teaching staff;

- the request from the enterprises of the military-industrial complex of Russia and industrial enterprises of the Omsk region for targeted training of students;
- the possibility of obtaining a labour profession;
- the opportunity to participate in international educational and cultural exchange programs;
- the effective system for professional orientation of schoolchildren.

The main key partners among the educational organizations are: Bauman University, Tomsk Polytechnic University, Moscow Aviation Institute, ITMO University, Polytechnic University of Milan, TU Wien. As well as leading scientific organizations are Omsk branch of Sobolev Institute of Mathematics SB RAS; JSC "Omsk Scientific-Research Institute of Instrument Engineering"; JSC Concern "Sea Underwater Weapon—Gidropribor"; Institute of Problems of Chemical Physics RAS; Space Research Institute RAS; Voevodsky Institute of Chemical Kinetics and Combustion SB RAS; Central Research Institute for Machine Building and others. The main employers and partners of the university from the professional sphere are IG "Polyot"—the branch of Khrunichev State Research and Production Space Center, A.M. Isayev Chemical Engineering Design Bureau, Central Aerohydrodynamic Institute, Salyut Machine-Building Production Association, Omsk Engine Design Bureau, JSC "High Technologies", JSC "Siberian devices and systems" and others.

The technical and technological orientation of the university forms the main list of educational programs and key areas for scientific research implemented in the field of technical, physical, mathematical and chemical sciences. The university has a number of world-class breakthroughs in the main areas of scientific specialization. 19 of 40 bachelor's degree programs implemented by the university meet the priority areas of modernization and technological development of the Russian economy. The university is the region leader in employer-sponsored training for enterprises and organizations among universities subordinate to the Ministry of Education and Science of the Russian Federation.

OmSTU is the main executor of the Strategy for socio-economic development of the Omsk region up to 2025 in terms of priority economic clusters staffing. The university has stable develop ties with the leading enterprises of the region: the work of 15 basic departments carrying out targeted training of specialists has been organized. It is a leader among the region universities in the volume of research work, including the interests of industrial partners. The university accumulates work in the field of youth policy, culture and tourism development of the Omsk region, participates in ensuring intercultural and interethnic dialogue. OmSTU has more than 16 thousand students, including about 3 thousand foreign students,being the leader by this indicator in the region. The teaching staff consists of more than 840 employees, including scientists, doctors and candidates of sciences.

The University is allied with educational, industrial, research organizations and institutions in Russia, Belarus, Kazakhstan, Germany, Italy, Austria, the Republic of Korea, etc.

OmSTU has been developing dynamically in recent years: new specialties and specializations, new computer classrooms, the number of students is twice the average number of state-run university students in Russia. The library at OmSTU is the largest of the Omsk university libraries, the library fund together with its branches amounts to 1 million 327 thousand publications. Internet technologies and other modern forms of working with readers are used actively in OmSTU.

3 Results of the analysis of educational programs

The evaluation criteria by which the indicators are checked, the maximum score for each criterion, the evaluation results [5] and the notes of the expert commission on the bachelor's educational program 24.03.01 "Rocket Systems and Astronautics" are shown in table 1.

Table 1. Evaluation results and notes of the expert commission on the program 24.03.01

Evaluation criterion	Maximum score	Evaluation result	Notes of the expert commission
Criterion 1. Passing a professional exam in the form of an independent qualification assessment by graduates of the professional educational program. Orientation of the planned program results to the requirements of professional standards	10	8	The degree of criterion 1 fulfillment was reduced by 20% because the independent assessment of graduates' qualification for compliance with the professional standards "Process Engineer for Assembly Operation in Aerospace Industry", and "Structural Engineer in Rocket Science" is not carried out
Criterion 2. Compliance of the planned results formulated in the educational program with professional standards in the rocket and space industry	6	5	There are no job descriptions, job responsibilities developed on the basis of the qualification requirements of employers. However, 31% of syllabuses are coordinated with the employer. The planned results of study of the educational program are limited only by two labor functions of the professional standard
Criterion 3. Compliance of curricula, syllabuses of academic subjects, courses, disciplines (modules), evaluation materials and procedures with the planned results of study of the educational program	14	11	It is necessary to update the Evaluation Tools in accordance with the recommendations of experts and to make additions to the regulations of the OmSTU quality management system, attaching the right for employers and other external experts to participate in the process of updating the educational program
Criterion 4. Human resources ensuring the formation, implementation and quality of the educational program	12	10	A small number of teaching staff (5.2%) involved in the implementation of the program have completed advanced training courses, retraining, and internships in specialized organizations. Insufficient number of teaching staff (10.4%) combines work in the educational organization with professional activity

Table 1. (Continue)

Evaluation criterion	Maximum score	Evaluation result	Notes of the expert commission
Criterion 5. Compliance of material and technical, information and communication resources and other ones influencing the quality of graduate training with the content of professional activity for which the graduate is trained	12	11	The basic department of the educational organization was transformed into a branch of the department on the basis of IG "Polyot", i.e. there was the downgrading of the department. When working with the electronic information system, it was revealed that the methodological guidelines for providing educational program activity have not been updated since 2015
Criterion 6. The presence of demand for the educational program, the demand for graduates by employers	14	11	To work out a system for monitoring the retention of the program graduates in the workplaces of enterprises after the first year of employment
Criterion 7. Participation of employers in the preparation and implementation of the educational program	12	11	It is necessary to reconsider the approaches of involving employers in the development of evaluation tools used in the educational program

Table 2. Evaluation results and notes of the expert commission on the program 24.04.01

Evaluation criterion	Maximum score	Evaluation result	Notes of the expert commission
Criterion 1. Passing a professional exam in the form of an independent qualification assessment by graduates of the professional educational program. Orientation of the planned program results to the requirements of professional standards	10	7	The degree of criterion 1 fulfillment was reduced by 30% because the independent assessment of graduates' qualifications for compliance with the professional standard "Structural Engineer in Rocket Science" is not carried out

Table 2. (Continue)

Evaluation criterion	Maximum score	Evaluation result	Notes of the expert commission
Criterion 2. Compliance of the planned results formulated in the educational program with professional standards in the rocket and space industry	6	5	The degree of criterion 2 fulfillment was reduced by 17% because there are no additional professional competencies as part of the planned results of study the educational program
Criterion 3. Compliance of curricula, syllabuses of academic subjects, courses, disciplines (modules), evaluation materials and procedures with the planned results of study of the educational program	14	11	The degree of criterion 3 fulfillment was reduced by 21.4% because the conformity of the evaluation procedures and tools used during the interim and final certification with the requirements established in the system of independent qualification assessment was not carried out
Criterion 4. Human resources ensuring the formation, implementation and quality of the educational program	12	11	The insufficient number of teaching staff (14%) have completed internship programs at enterprises of the rocket and space industry. It is necessary to improve the mechanism for monitoring employers' satisfaction with the quality of graduate training
Criterion 5. Compliance of material and technical, information and communication resources and other ones influencing the quality of graduate training with the content of professional activity for which the graduate is trained	12	11	The basic department was transformed into a branch of the department on the basis of IG "Polyot", i.e. there was the downgrading of the department. When working with the electronic information system, it was revealed that the methodological guidelines for providing educational program activities have not been updated since 2015
Criterion 6. The presence of demand for the educational program, the demand for graduates by employers	14	11	Monitoring the retention of graduates occurs through non-formalized sources. The educational organization needs to strengthen control over graduates at enterprises

Table 2. (Continue)

Evaluation criterion	Maximum score	Evaluation result	Notes of the expert commission
Criterion 7. Participation of employers in the preparation and implementation of the educational program	12	11	In the disciplines of the program, taught by representatives of employers, evaluation materials are developed by employers. The rest of the evaluation tools was developed by the teachers of the department

The evaluation criteria by which the indicators are checked, the maximum score for each criterion, the evaluation results and the notes of the expert commission on the master's degree program 24.04.01 "Rocket Systems and Astronautics" are shown in table 2.

Experts noted the following strengths of the educational programs "Rocket Systems and Astronautics" 24.03.01 and 24.04.01.

1. Practical orientation in comparison with competitor programs. The results of study of educational programs are aimed at improving the competitiveness of the Omsk region, making OmSTU a driver for the implementation of the cluster model for industrial development in the region and the regional flagship university.
2. The practical orientation of the principal educational program is confirmed by a large proportion of practice-oriented materials in the syllabuses of disciplines and evaluation tools, discussed and approved with employers.
3. High demand for graduates.
4. The high publication activity of the teaching staff ensures better formation and implementation of programs.
5. The OmSTU quality management system of education with the established procedure for managing the measurement of customer satisfaction with the university's services allows improving the quality of education.
6. The rating system operating at OmSTU for material rewards of the most active teachers in creative and scientific terms significantly improves the quality of educational programs.
7. The high state level of the OmSTU material base ensures high-quality fulfillment of all requirements for the educational process.
8. The current system of motivating promising young teachers contributes to their retention in OmSTU.

There were no significant deficiencies in the bachelor's degree programs 24.03.01 "Rocket Systems and Astronautics" and the master's degree 24.04.01 "Rocket Systems and Astronautics" as a result of the inspection by the commission.

On June 27, 2019, at the meeting of the Accreditation Council for Professional and Public Accreditation of professional educational programs in the field of rocket technology and space activity, it was unanimously decided to accredit and issue certificates for the two abovementioned OmSTU programs until June 2024.

Based on the results of the comments made by the commission, the action plan was developed to eliminate them, which was successfully implemented in 2019–2020 academic year.

References

[1] Federal Law No. 273-FZ of 29.12.2012, "On Education in the Russian Federation" (the latest edition), URL http://www.consultant.ru/document/cons_doc_LAW_140174/

[2] A.A. Factorovich, *Professional and public accreditation of educational programs as a tool for quality management of vocational education*, Additional professional education in the country and the world J., **6–7(20–21)**, 1 (2015)

[3] Questionnaire for self-examination during professional and public accreditation of educational programs in the field of rocket technology and space activity, URL https://profaccred.ru/documents

[4] A.B. Yakovlev, E.V. Yakovleva, *Innovation-oriented aerospace education at Omsk State Technical University*, in Proceedings of the All-Russian Scientific Conference "Cosmonautics: science and education", 17 October 2019, Blagoveshchensk, Russia (2019)

[5] Methodology for professional and public accreditation of educational programs in the field of rocket technology and space activities, URL https://profaccred.ru/documents

AUTHOR INDEX